Digital Electronics Laboratory Experiments
Using the
Xilinx® XC95108™ CPLD
with
Xilinx® Foundation™
Design and Simulation Software

James W. Stewart
Chao-Ying Wang
DeVry Institute

Prentice
Hall

Upper Saddle River, New Jersey
Columbus, Ohio

Vice President and Publisher: Dave Garza
Editor in Chief: Stephen Helba
Acquisitions Editor: Scott J. Sambucci
Production Editor: Rex Davidson
Copyeditor: Carol Mohr
Design Coordinator: Robin G. Chukes
Cover Designer: Thomas Borah
Cover Art: Marjory Dressler
Production Manager: Pat Tonneman
Marketing Manager: Ben Leonard

The book was set in Times New Roman by Chao-Ying Wang and James W. Stewart. It was printed and bound by Victor Graphics, Inc. The cover was printed by Victor Graphics, Inc.

10 9 8 7 6 5 4 3

ISBN 0-13-088192-9

CONTENTS

PREFACE

Up until now, almost all lab manuals for introductory digital courses at the Electronics Engineering Technology (EET) or Electronics and Computer Technology (ECT) level have been written around the use of TTL chips. One reason for not moving to CPLDs was the lack of a suitable target board at a reasonable price. But now several vendors are supplying such boards. The switches and LEDs of the target board are used to supply the input-output functions. Another reason, we believe, was the perception that a steep learning curve had to be climbed in order to use the software tools. The reality is that it's not as steep as it seems. This manual is an attempt to show the flatness of that curve by providing a set of hands-on lab jobs with step-by-step instructions on using the design software. The lab jobs are based on the Xilinx XC95108 CPLD and use the student version of the Xilinx® Foundation™ series software.

The manual is organized in two sections. The first, shorter section is a set of labs using TTL chips. This allows the students to build a few simple circuits immediately. Also, TTL is not totally gone; it is still used as "glue logic" in some applications. So it is still worthwhile for students to get their hands on the chips.

The second section, on using the CPLD, is the bulk of the manual. The first few labs in the CPLD section explore basic gates and Boolean algebra. We then move on to combinatorial circuits, including adders, multiplexers, encoders, and decoders. Next we explore latches and flip-flops, followed by counters and registers. Appendices include data for the XC95108 as well as documentation for two target boards. The appendices also include two tutorials and a glossary of terms for reference.

Selecting a target board is an important task for the instructor using this manual. The first decision is whether to build or buy. If the decision is to build, the board described in the appendix of Dave Van den Bout's book *The Practical Xilinx® Designer Lab Book* is a good example. If the decision is to buy, two possibilities are the XS95™ / XStend™ board combination from XESS® Corporation and the PLDT-1™ board from RSR® Electronics. The XESS board set is more advanced and supports mouse, VGA, and CODEC interfaces as well as switches, LEDs, and displays. It has an on-board 8051 microcontroller. The RSR board is a basic prototyping board with switches, LEDs, a 7-segment display, and connectors for ribbon cables. In writing this manual, we thought it would be useful to refer to a specific target board in order to avoid vagueness. So, many of the labs in this manual refer to the PLDT-1™ board. However, the labs can be implemented on any target board using the same CPLD device.

We wish to thank the following people for their support and help on this project: Amin Karim of DeVry; Eric Addeo, Raul Lasluisa (student), and our colleagues at DeVry Institute of New Jersey; Patrick Kane and Alfred Rodriguez of Xilinx®; Ajit Gulati and Robert Wichiciel of RSR® Incorporated; Dave Van den Bout of XESS® Corporation; Professor Muhammad Mazidi of DeVry–Dallas; and Scott Sambucci and Toni Payne of Prentice Hall.

Xilinx® is a registered trademark of Xilinx®, Inc. ISP™ is a trademark of Lattice® Semiconductor Corporation. Other product and company names mentioned are trademarks or trade names of their respective companies.

VENDORS

The PLDT-1™ CPLD prototyping board, as well as electronic components, are available from:

> **Electronix Express / RSR Electronics**
> **365 Blair Road**
> **Avenel, NJ   07001**
> **(732) 381-8777**
> **www.elexp.com**

The XS95-108™ CPLD prototyping Board and the XST-1 XStend™ I/O prototyping extender board are available from:

> **XESS Corporation**
> **2608 Sweetgum Drive**
> **Apex NC   27502**
> **(800) 549-9377**
> **www.xess.com**

INTRODUCTION

While the principles of digital design have remained constant, the technology in which such designs are implemented has changed rapidly. From transistors in the 1950s to small-scale integrated circuits (ICs) in the 1960s to large-scale ICs in the 1970s, to the sophisticated programmable ICs of today

Early digital ICs were made as standard building blocks that had to be interconnected by copper traces on a circuit board. By the 1980s, the blocks could be mounted on one chip, and the interconnection done by "burning" a design into it. Those were the early programmable logic devices (PLDs). They were one-time programmable (OTP), and required a piece of equipment called a device programmer to do the "burning". Later came devices that could be erased electrically and reprogrammed.

Today, ICs such as the venerable 7400 family of TTL chip are almost obsolete. Modern digital designs use large-scale programmable ICs such as field-programmable gate arrays (FPGAs) and complex programmable logic devices (CPLDs). Internally, FPGAs implement logic functions using look-up tables in memory blocks while CPLDs use sum-of-product terms configured from arrays of gates. Both FPGAs and CPLDs also contain flip-flops.

To program FPGAs and CPLDs, special computer-aided design (CAD) software is required that allows the user to enter a design on a PC, check it for validity, simulate its performance, and then download it into the target chip. A design can be entered using schematic capture software to draw a logic diagram of interconnected gate symbols. Or a design may be entered using a text editor to create a file containing a set of Boolean equations written in a hardware description language (HDL). A common HDL is ABEL (Advanced Boolean Expression Language). Either way, the CAD software compiles the design into the low-level commands needed to configure the device.

Once a design has been entered, the software can test the functional and timing performance of it with simulation. After simulation, the design can be downloaded into the CPLD device while the device is in the circuit. The CPLD will hold the design in memory even without power being applied. But the device can be erased and reprogrammed with a new design over and over again.

Prototyping digital circuits using CAD software on a programmable logic device may seem a lot more complicated than just wiring up chips on a breadboard. Looking at a computer screen full of pop-up windows, and unfamiliar terms (JTAG?) and procedures (Integrity test?) can be intimidating. You may have felt the same way about learning how to ride a bike or drive a car. But once you had a chance to practice, the unfamiliar became familiar and you found that you could do it easily. And so it will be here. This manual has been designed to guide you step-by-step through the process of creating and testing a design in software, and down-loading it to a CPLD. When you are finished with this manual, it will be no more intimidating than using a TTL manual and a soldering iron (but without the solder burns).

The lab experiments in this manual have been written for the Xilinx XC95108 CPLD device using the Xilinx Foundation software. But the skills and concepts learned can be transferred to other similar devices using similar design software.

Part One

TTL EXPERIMENTS

TTL EXPERIMENT 1:
Basic Gates: NOT, AND, NAND, OR, NOR

OBJECTIVES:

- Examine logical operations.
- Examine basic logic circuits.
- Examine the operation of a standard small-scale integrated TTL device.

MATERIALS:

- 74LS00 Quad-2 NAND gate IC in a 14-pin DIP.
- Solderless breadboard.
- Roll of insulated, solid 22 Ga. hook-up wire.
- Wire cutters, wire strippers, long-nosed pliers.
- Digital Voltmeter.

DISCUSSION:

Logical Operations
The operation of computers and all other digital equipment is based on a few logical concepts and operations. They are so simple that it is hard to believe so much can be derived from them. The concepts include truth-values, binary numbers, and Boolean algebra. The operations are: AND, OR, and NOT which three can be combined to form two others just as simple: NAND and NOR.

Truth-Values & Binary Numbers
Many statements can be evaluated as being either true of false. Some simple examples are:
- statement A: John is six feet tall.
- statement B: Joe is six feet tall.
- statement C: John and Joe are both six feet tall.

If statements A and B are true, then C must be true. Obviously if either A were false or B were false, then C would have to be false. Let the digit 1 represent "is true" and let the digit 0 represent "is false". Then we can summarize the statement possibilities in a table:

A	B		C
0	0		0
0	1		0
1	0		0
1	1		1

Table L1.1 This Is a Truth Table

3

AND

The truth table above represents the logical AND operation. We can write the AND operation as an equation:

$$C = A \bullet B \quad \text{(or} \quad C = AB \text{ where the dot is assumed)}$$

where the equal sign (=) represents the words "is true if" and the dot (•) represents the word "and". So the equation means:

Statement C *is true if* statement A is true *AND* statement B is true. Otherwise, statement C is false.

Because we are dealing with things that can be true or false only, we represent them by binary numbers, which can be 0 or 1 only. Let's look at the electrical circuit in figure L1.1 below. It's composed of a battery, two switches, and a light bulb. Obviously, for the light bulb labeled L to light, switch A and switch B must both be closed. Use binary 1 to mean "switch closed" and binary 0 to mean switch open. For the light bulb, use binary 1 to mean "on" and binary 0 to mean "off". Then we can write the Boolean equation $L = A \bullet B$, and the truth table would be the same as Table L1.1 above.

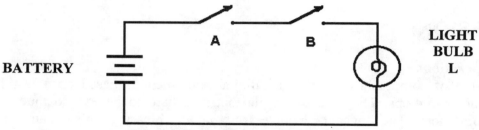

Figure L1.1 Electrical AND circuit.

OR

Now let's look at the circuit in figure L1.2 below. This time, for the light bulb to be lit, either switch A *OR* switch B must be closed, or both can be closed.

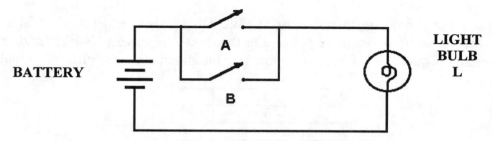

Figure L1.2 Electrical OR circuit.

4

The truth table for the OR operation is:

A	B		L
0	0		0
0	1		1
1	0		1
1	1		1

Table L1.2 Truth Table for OR Operation.

NOT

The last fundamental operation is NOT, also called *logical inversion*. NOT means that one thing is logically the opposite of the other. For example, let "Mary is taller than Jane" be statement A, and let "Jane is as tall as, or taller than Mary" be statement B. If one statement is true, than the other must be false. We can write that as a Boolean equation:

$$B = \overline{A}$$

The truth table is:

A		B
0		1
1		0

Table L1.3 Truth Table for NOT Operation.

GATES

In digital electronic equipment, logical operations are carried out by circuits called *gates*. The output of a gate is the logical combination of the inputs. Inputs and outputs are represented by voltages: +5 (or +3.3) Volts for a logical 1 and 0 Volts (ground) for a logical 0. Physically, gates come in integrated circuit "chips" that are soldered to circuit boards. They are usually rectangular with metal pins along two or more edges. Figure L1.3 shows the outline of the 74LS00 TTL (transistor-transistor logic) chip.

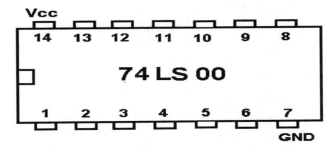

Figure L1.3 74LS00 Integrated logic circuit.

GATES: SYMBOLS & OPERATIONS

1) The AND Gate

Symbol	Boolean Equation	Truth Table

Symbol: A, B → X

Boolean Equation: $X = A\,B$

Truth Table:

Inputs		Output
A	**B**	**X**
0	0	0
0	1	0
1	0	0
1	1	1

2) The OR Gate

Symbol: A, B → Y

Boolean Equation: $Y = A + B$

Truth Table:

Inputs		Output
A	**B**	**Y**
0	0	0
0	1	1
1	0	1
1	1	1

3) The INVERTER (NOT Gate)

Symbol: A → X

Boolean Equation: $X = \overline{A}$

Truth Table:

Input	Output
A	**X**
0	1
1	0

If we combine an AND gate with an inverter,

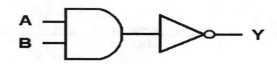

we get the NAND gate on the next page.

4) **The NAND Gate**

Symbol	Boolean Equation	Truth Table

	Inputs		Output
A	**B**		**Y**
0	0		1
0	1		1
1	0		1
1	1		0

A
B ⊐D⊳o– Y

$$Y = \overline{A B}$$

If we combine the OR gate with an inverter,

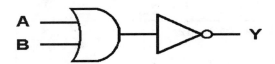

we get the NOR gate.

5) **The NOR Gate**

Symbol	Boolean Equation	Truth Table

	Inputs		Output
A	**B**		**Z**
0	0		1
0	1		0
1	0		0
1	1		0

A
B ⊐D⊳o– Z

$$Z = \overline{A + B}$$

PROCEDURE:

1)   Insert the 74LS00 chip onto the solderless breadboard so it straddles the center line as shown in figure L1.4 below.

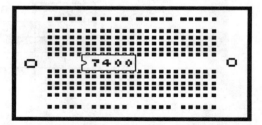

Figure L1.4 Solderless breadboard.

Figure L1.5 shows the pin assignments, or *pin-out* for the chip.

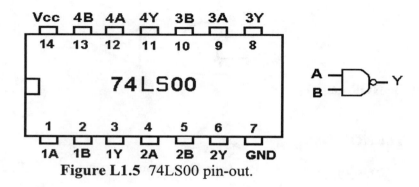

Figure L1.5 74LS00 pin-out.

2)   Cut three 4" lengths of hook-up wire and strip both ends of each. Insert one of the wires into the solderless breadboard to connect to the input on pin 1 of the integrated circuit chip. Insert one end of the second wire to connect to pin 2 of the IC. Insert one end of the third wire to connect to pin 3 of the IC chip.

3)   Cut two lengths of wire long enough to connect the solderless breadboard to the 5 Volt power supply. Connect one end of one wire to the ground terminal on the power supply. Connect the other end to pin 7 of the chip. Connect one end of the other wire to the +5 Volt terminal on the power supply. Connect the other end to pin 14 of the chip.

Checked by _____ Date _____

4) Use a digital voltmeter to measure the voltage on the pin 3 output of the chip. By connecting the two input wires to ground (logic 0) or to +5 Volts (logic 1), complete the truth table:

PIN 1	PIN 2		PIN 3
0	0		
0	1		
1	0		
1	1		

Do your results verify a NAND operation?

Checked by _____ Date _____

5) Next we will use a 2-input NAND gate to work as an inverter. Using pins 4, 5, and 6, tie the two inputs together with pieces of hook-up wire as shown in figure L1.6 below.

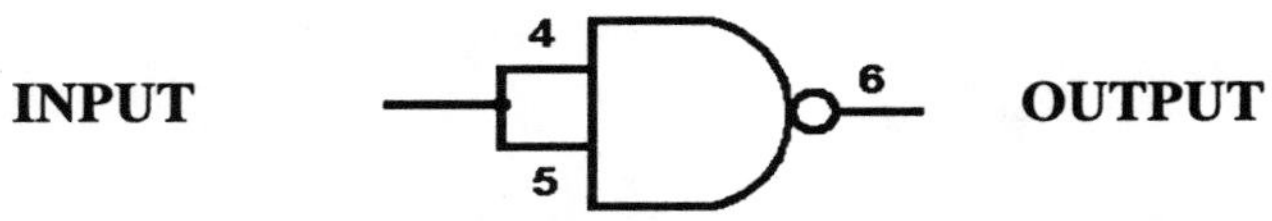

Figure L1.6 NAND used as an inverter.

Using +5 V = 1 and GND = 0, fill in the truth table and verify that the circuit is an inverter:

INPUT		OUTPUT
0		
1		

Checked by _____ Date _____

9

6)   Next we will implement an AND gate using two NAND gates as shown in figure L1.7 below.

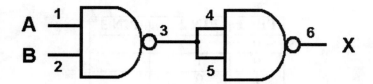

Figure L1.7 AND gate using two NAND gates.

Cut and strip enough hook-up wire to build the circuit of figure L1.7 on the solderless breadboard. Using +5 V = 1 and GND = 0, fill in the truth table and verify that the circuit is an AND gate.

A	B		X
0	0		
0	1		
1	0		
1	1		

Checked by _____ Date _____

7) Next we will implement an OR gate using three NAND gates as shown in figure L1.8 below.

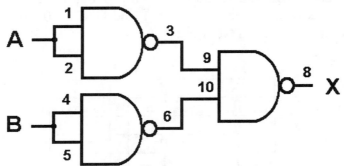

Figure L1.8 OR gate using three NAND gates.

Cut and strip enough hook-up wire to build the circuit of figure L1.8 on the solderless breadboard. Using +5 V = 1 and GND = 0, fill in the truth table and verify that the circuit is an OR gate.

A	B		X
0	0		
0	1		
1	0		
1	1		

Checked by _____ Date _____

8) Design and build a NOR gate.

In the space below, draw a circuit that implements a NOR gate using four NAND gates. Cut and strip enough hook-up wire to build the circuit you drew on the solderless breadboard. Using +5 V = 1 and GND = 0, fill in the truth table and verify that the circuit is an NOR gate.

A	B		X
0	0		
0	1		
1	0		
1	1		

Checked by _____ Date _____

NAME _____ DATE _____

QUESTIONS:

1)   Explain why a NAND gate could be called a "universal gate".

2)   Using a quad-2 NOR gate digital IC chip, could you implement an inverter, an AND gate, and an OR gate?   Draw the circuits here.

NAME _____ **DATE** _____

TTL EXPERIMENT 2:
A Gate Circuit using the 74LS00

OBJECTIVES:

- Build a simple combinatorial circuit using a TTL gate chip.
- Analyze the expected behavior of a simple digital circuit.
- Take data to verify performance of a digital circuit.

MATERIALS:

- Solderless breadboard.
- 74LS00 TTL Quad-2 integrated circuit.
- Roll of solid 22 Ga. insulated hook-up wire.
- Wire cutters, wire strippers, long-nosed pliers.
- Digital voltmeter.
- A target board with at least 4 switches and 3 LEDs.

DISCUSSION:

In the previous experiment we defined the behavior of the logical operations AND, OR, NOT, NAND, and NOR. We also implemented those operations using NAND gates in a 74LS00 TTL chip. In this experiment, we will extend those basic concepts to analyze the performance of several gates connected together into a *combinational* (or *combinatorial*) logic circuit.

We can analyze large combinational circuits with a "bottom-up" approach. Starting at the inputs, calculate the outputs of the first layer of gates. Using those outputs as inputs to the next layer of gates, calculate the outputs of the second layer. Continue until you get to the final output.

This is an example of the "divide-and-conquer" strategy. We divide a large, complex problem into a collection of small, simple problems. We solve the simple ones and thereby solve the complex one.

15

PROCEDURE:

1) Use what you learned from the previous experiment to get equations for A, B, and Q from the following circuit:

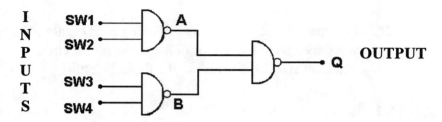

Figure L2.1 A digital circuit.

A = _____

B = _____

Q = _____

2) Using the equation, fill in for A, B, and Q in table L2.1 on the next page.

3) Cut and strip enough wire to build the circuit on the solderless breadboard. Refer to figures L2.2 and L2.3 below. Connect to a 5 Volt power supply.

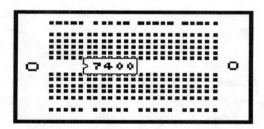

Figure L2.2 A solderless breadboard.

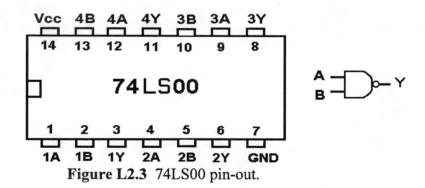

Figure L2.3 74LS00 pin-out.

16

4) Cut and strip enough wire to connect the circuit inputs to four switches on your target board and to connect the A, B, and Q outputs to LEDs. Assuming a switch set at H is a logic 1 and a lit LED is a logic 1, fill in the column labeled **LED** in Table L2.1 below.

SW 1	SW 2	SW 3	SW 4	A	B	Q	LED
0	0	0	0				
0	0	0	1				
0	0	1	0				
0	0	1	1				
0	1	0	0				
0	1	0	1				
0	1	1	0				
0	1	1	1				
1	0	0	0				
1	0	0	1				
1	0	1	0				
1	0	1	1				
1	1	0	0				
1	1	0	1				
1	1	1	0				
1	1	1	1				

Table L2.1 Truth Table for Digital Circuit of figure L2.1

5) Does the LED column match the Q column? Do your results verify your equation?

Checked by _____ Date _____

6) Modify the digital circuit: Use the last gate in the 7400 as an inverter, and insert it between SW1 and the input to the NAND gate.

7) Recalculate the Boolean equations, and use them to fill in the A, B, and Q columns in table L2.2 below.

A = _____

B = _____

Q = _____

SW 1	SW 2	SW 3	SW 4	A	B	Q	LED
0	0	0	0				
0	0	0	1				
0	0	1	0				
0	0	1	1				
0	1	0	0				
0	1	0	1				
0	1	1	0				
0	1	1	1				
1	0	0	0				
1	0	0	1				
1	0	1	0				
1	0	1	1				
1	1	0	0				
1	1	0	1				
1	1	1	0				
1	1	1	1				

Table L2.2 Truth Table for Modified Digital Circuit

6) As before, use the switches and LEDs of the target board to fill in the LED column. Does the LED column match the Q column? Do your results verify your equation?

Checked by _____ Date _____

QUESTIONS:

1)   Redraw the circuit of figure L2.1 by replacing the NAND gates with NOR gates.

2)   Write the equation for Q for the new circuit.

3)   Make a truth table similar to Table L2.1 and fill it in.

NAME _____ **DATE** _____

TTL EXPERIMENT 3:
The Cross-Coupled Latch & Switch De-Bouncing

OBJECTIVES:

- Define and describe an S-R latch.
- Experimentally verify the performance of a latch.
- Use a latch to de-bounce a switch.

MATERIALS:

- Solderless breadboard.
- 74LS00 TTL Quad-2 NAND integrated circuit.
- Two 10k resistors, 1/4 W.
- Roll of solid 22 Ga. insulated hook-up wire.
- Wire cutters, wire strippers, long-nosed pliers.
- Digital counter, oscilloscope.
- A target board with at least 2 switches and 2 LEDs

DISCUSSION:

In the previous experiments we examined logic gates and how they can be combined into a small combinational digital circuit. A characteristic of combinational circuits is that, as soon as the inputs are removed, the outputs return to their original state. In other words, simple combinational circuits have no memory. But memory, the ability to store a 0 or 1 state after inputs are removed, is necessary to store programs and data in a computer.

In this experiment we will examine the S-R latch. A latch is a simple circuit capable of storing one bit of information for as long as the power is on. The letters S-R stand for Set and Reset. The Set input will force the latch to store a 1, while the Reset input will cause it to store a 0. If a latch is set, then even if the inputs are removed, it remains set until it is reset. We can make a latch by cross-coupling two NAND gates:

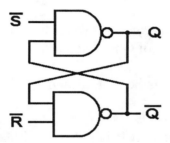

Figure L3.1 Cross-coupled NAND latch.

21

The inputs on Figure L3.1 are labeled $\overline{S}$ and $\overline{R}$ because they are "active low", meaning that a logic level of 0 will *assert* the input. Remember that inverting a 0 gives a 1, so putting 0 on an active low input is the same as putting 1 on an "active high" input.

To analyze this latch, let's assume that both inputs are high (i.e., not active). What will the outputs be? Let's assume that Q is low and "Q-bar" is high as shown in Figure L3.2.

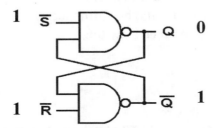

Figure L3.2 Q initially low.

Then the 0 output from the upper gate drives an input on the lower gate. Since it's a NAND gate, a 0 on an input forces the output of the lower gate to 1. That 1 output is then applied to one of the upper gate's inputs. Since both inputs of the upper NAND gate are 1s, the output will be low. Thus our assumption about the outputs was correct.

Instead of Q being initially low, let's assume it was initially high as shown in Figure L3.3 below. Can that work? You can work through the analysis as we did above. Or you might notice that the circuit is symmetrical with respect to inputs and outputs. That means if we swap the S and R inputs with each other, and also swap the Q outputs with each other, the circuit will remain the same. By either reasoning, you will find that the latch could have started out with Q high.

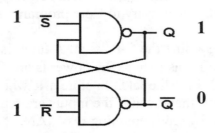

Figure L3.3 Q initially high.

The latch is said to be in its *holding state* when both inputs are inactive (in this case, that's both inputs high). The value of Q is "latched".

If $\overline{S}$ is asserted with a 0 while $\overline{R}$ is held at 1, then Q will go high regardless of what it was before asserting the input.

Likewise (remember the symmetry) if $\overline{R}$ is asserted with a 0 while $\overline{S}$ is held at 1, then $\overline{Q}$ will go high regardless of what it was before asserting the input.

So the way a latch works is that it is either set or reset by asserting either the S or the R input. When both inputs return to the inactive level, the Q output will stay either set or reset. What happens when both inputs are asserted and then return to the inactive level? The answer is that it's *indeterminate*. That means Q could be 1 or 0, we don't know. Having indeterminate states in a digital circuit usually is not a good thing.

Table L3.1 summarizes the operation of a latch.

HAPPENS FIRST	HAPPENS SECOND	HAPPENS THIRD		RESULT ON OUTPUT
S inactive R inactive	S active R inactive	S inactive R inactive		$Q = 1$
S inactive R inactive	S inactive R active	S inactive R inactive		$Q = 0$
S inactive R inactive	S active R active	S inactive R inactive		$Q = ?$

Table L3.1 Operation of an S-R Latch

Figure L3.4 below shows the operation of a latch in the form of a *timing diagram*. It contains the same information as is in Table L3.1 above. A timing diagram is what you could see when looking at the inputs and output with an oscilloscope.

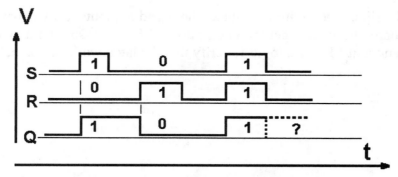

Figure L3.4 Timing diagram for an S-R latch.

One common use for an S-R latch is to *de-bounce* mechanical switches. When a switch is activated and the metal contacts come together, the contacts will bounce off each other a few times before coming to rest in a closed position. The "bounce-time" is on the order of 10 milliseconds. But digital circuits can operate in nanoseconds, so they see the bouncing as multiple switch closures. That's a problem if the circuit is trying to count how many times a switch is pressed. We see how to fix that problem below.

PROCEDURE:

Latch Operation

1) Cut and strip enough wire to build the latch circuit on the solderless breadboard. Refer to Figures L3.5 and L3.6 below. Connect to a 5 Volt power supply.

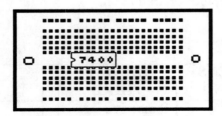

Figure L3.5 Solderless breadboard with IC.

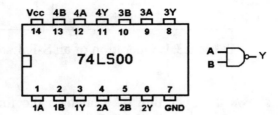

Figure L3.6 74LS00 pin-out.

2) Cut and strip enough wire to connect the $\overline{S}$ and $\overline{R}$ inputs to two switches on your target board and to connect the Q outputs to LEDs. Assuming a switch at H is a logic 1 and a lit LED is a logic 1, verify that the latch works as expected.

Checked by _____ Date _____

24

Switch De-Bouncing

Build the circuit of Figure L3.7 shown below using 10k resistors.

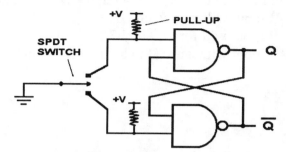

Figure L3.7 Switch de-bounce circuit.

Use a toggle switch or push-button switch from your target board. If you have access to an electronic counter, attach it to the input of the upper NAND gate at the point where the pull-up resistor is connected. Activate the switch and note how many counts the counter registers. Then attach the counter to the Q output and activate the switch. Verify that you get one count per activation.

If you have access to an oscilloscope (scope), attach it to the input of the upper NAND gate at the point where the pull-up resistor is connected. Put the scope into auto trigger mode, and set the sweep speed for 1 ms/div. Activate the switch repeatedly. You should be able to see the trace make several high/low transitions before stabilizing. You may have to adjust the sweep speed.

Then attach the scope to the Q output and activate the switch. Verify that you see one transition between high and low.

Checked by _____ Date _____

NAME _____ DATE _____

QUESTIONS:

1) Redraw the circuit of Figure L3.3 by replacing the NAND gates with NOR gates.
    a) What S-R input will now latch Q?
    b) What S-R input will force Q high?
    c) What S-R input will cause Q to be indeterminate?

2) For the circuit of question 1, draw a timing diagram similar to Figure L3.4 above.

3) Can you use the NOR latch to de-bounce a switch as you could with the NAND latch? Explain your answer.

Part Two

CPLD EXPERIMENTS

CPLD EXPERIMENT 1:

AND Gates & OR Gates

OBJECTIVES:

- Create new projects and save schematic designs in Xilinx Foundation Software.
- Download JEDEC files to the target board.
- Demonstrate the characteristics of AND and OR gates.
- Develop truth tables for AND and OR gates.

MATERIALS:

- Xilinx Foundation Software, student or professional edition V1.5 or higher.
- IBM or compatible computer with Pentium processor or equivalent, 64 M-byte RAM or more, and 3 G-byte or larger hard drive.
  NOTE: Read Appendix E on parallel port modes for a PC.
- PLDT-1 board by RSR Electronics Inc., XS95 (V1.2) board by XESS Corp., or a similar board with an XC95108 device.

DISCUSSION:

AND and OR gates are hardware implementations of the two fundamental Boolean operations: logical multiplication (AND) and logical addition (OR). They are represented on a diagram by gate symbols, and their characteristics can be described by truth tables, Boolean equations, and timing waveforms. Both AND and OR gates can have two or more inputs, but only one output. Although we use two inputs in our discussion, the principles apply to more than two inputs.

Gate Characteristics

1) **The AND Gate**

	Symbol	Boolean Equation	Truth Table

<table>
<tr><td rowspan="2" colspan="3">Symbol</td><td rowspan="2">Boolean Equation</td><td colspan="3">Truth Table</td></tr>
</table>

Symbol   Boolean Equation   Truth Table

$$X = A\,B$$

Inputs		Output
A	**B**	**X**
0	0	0
0	1	0
1	0	0
1	1	1

The behavior of an AND gate can be summarized as follows: The output is HIGH (true or logic 1) only when *all* the inputs are HIGH. If any of the inputs is LOW (false or logic 0), the output will be LOW.

2) **The OR Gate**

Symbol   Boolean Equation   Truth Table

$$Y = A + B$$

Inputs		Output
A	**B**	**Y**
0	0	0
0	1	1
1	0	1
1	1	1

As seen from the above truth table, the output is LOW only when *all* the inputs are LOW. If any of the inputs is HIGH, then the output is HIGH.

For both AND gates and OR gates, the output goes HIGH when the required HIGH logic levels are applied to the input. We describe this by saying that AND gates and OR gates have active-high inputs and active-high outputs.

PROCEDURE:

1) Create a directory called *Xilabs* on your C drive (the files will be too big for a floppy). You will save all your experiment files in this directory.

2) Start the Software by using your mouse to click on the following: **Start → Programs → Xilinx Foundation Series → Xilinx Foundation Project Manager** on your Windows 95/NT/98. You can also start by double clicking the **Project Manager** Icon on your Window.

3) In the **Getting Started** window (Figure 1.1), choose **Create a New Project** and click on the **OK** button.

Figure 1.1 Getting Started window.

4) When the **New Project** window appears, type *AND_OR3* in the **Name** editor box as your project file name. In the **Directory** editor box, either type or browse to get *Xilabs*. Choose **Type** as *F2.1i*. Click on **Schematic**. Make sure you select the **Family, Part,** and **Speed** to be *XC9500, 95108PC84* and *20*, respectively. Then, click On the **OK** button.

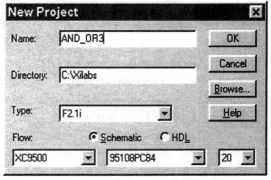

Figure 1.2 New Project window for project *AND_OR3*.

5) The new project is created and you should see the window shown in Figure 1.3.

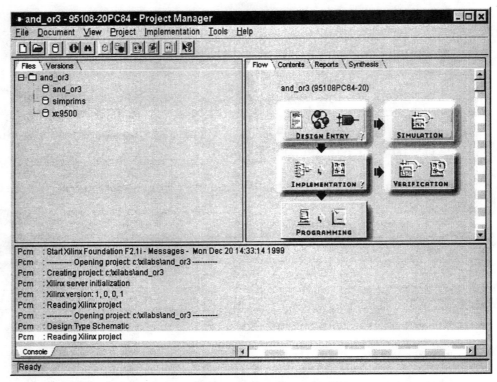

Figure 1.3 Project Manager window for project *AND_OR3*.

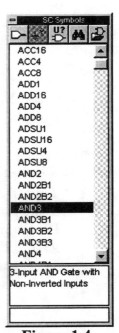

6) Click the **Schematic Editor** button ▮▯ on the **Design Entry** tab [DESIGN ENTRY ?] in the right pane to get a blank **Schematic Editor** window.

7) Click on the logic symbol icon ⟆ on the tool bar to open **SC Symbols** window. Move the glide bar on **SC Symbols** window to AND3. The message box indicates: "3-input AND Gate with Non-Inverted Inputs" as shown in **SC Symbols** window (Figure 1.4). Click your mouse on AND3, and then click on where you want the symbol to be in the **Schematic Editor** window. Get and place the OR3 gate symbol similarly. Your **Schematic Editor** window should resemble Figure 1.5 below.

Figure 1.4
SC Symbols window.

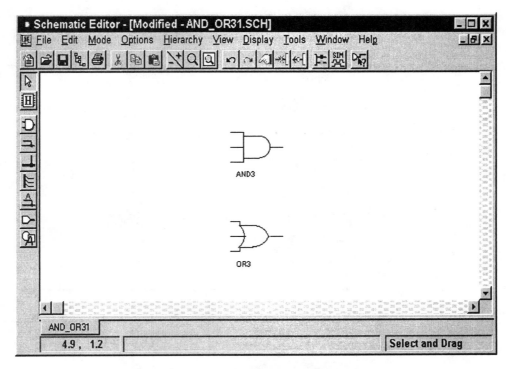

Figure 1.5 Schematic Editor window after step 7.

8)   Save this schematic file by using your mouse to choose **File → Save As.** In the **Save As** window, edit the **File name** box to *AND_OR31.SCH* and click on the **OK** button to save the file to the AND_OR3 subdirectory.

9)   In Tutorial 1 (Appendix C) we discuss how to add input buffers (IBUF) and output buffers (OBUF), and how to obtain I/O terminals. Add all necessary IBUFs, OBUFs, and I/O terminals. Then wire all the connections. Your **Schematic Editor** window should now look like Figure 1.6 below.

10)   Refer to Tables T1.3 and T1.4 in Tutorial I to see how to assign the six inputs A, B, C, and D, E, F to pins 6, 7, 11, and 71, 66, 70, respectively. This means that we are using toggle switches 1, 2, 3, 6, 7, and 8 on the PLDT-1 board. Assign LED 1 and LED 8 to display the outputs X and Y, that is pin 44 = X and pin 35 = Y. To assign pin 6 to input A, double click the buffer for signal A. In the **Symbol Properties** window, type *loc* (for locked) in the **Name** editor box and *p6* in the **Description** editor box. Then, click on the **Add** button followed by the **Move** button. Finish all the pin assignments in the same way. Your Schematic Editor should be similar to Figure 1.7 below.

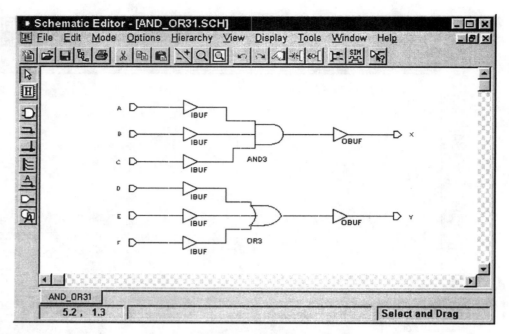

Figure 1.6 Schematic Editor window after step 9.

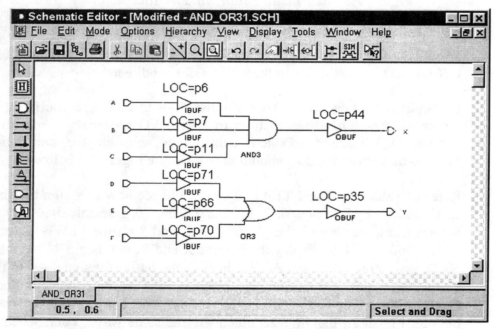

Figure 1.7 Schematic Editor window after step 10.

11)   You have finished the schematic drawing for the project AND_OR3 using Xilinx Foundation Series. Save your schematic by choosing **File → Save**.

12)   Choose **Option → Create Netlist** on the menu bar. The message should say "Netlist created successfully". Click on the **OK** button.

13) Choose **Option** → **Integrity Test**. The message "Integrity test passed successfully" should appear. Click on the **OK** button to go back to the **Schematic Editor** window.

14) Select **Option** → **Export Netlist**. Choose *Edit 200 [*.EDN]* for the **Files of type.** Click on the **Open** button to start the exporting process.

15) You have now completed your AND_OR3 design in the **Schematic Editor**. Return to the **Project Manager** window by selecting **File** → **Exit**. Verify that there is a green check mark on the **Design Entry** button.

16) Click on the implementation button to compile your design. Make sure the **Device, Speed, New version name,** and **New revision name** edit boxes have *95108PC84, 20, ver1,* and *rev1*, respectively. You need to stay consistent with the **Device** and **Speed** you chose at the beginning of design. Click on the **Run** button.

17) With no errors, the implementation process will move to the last stage: **Bitstream**. Any error, such as a disconnected component, will be detected during the second stage and the process will be terminated with the message "Implementation completed with errors". Even with errors, you will have gotten through the "Create Netlist", "Integrate Test" and "Export Netlist" in steps 12 to 14. Try disconnecting a wire and repeating the process to observe what happens. Then connect it back and repeat from step 12. If there is no error, you will see a window as shown in Figure 1.8. Select **File** → **Close** to go back to **Project Manager** window and verify that a green check mark appears on the **Implementation** button.

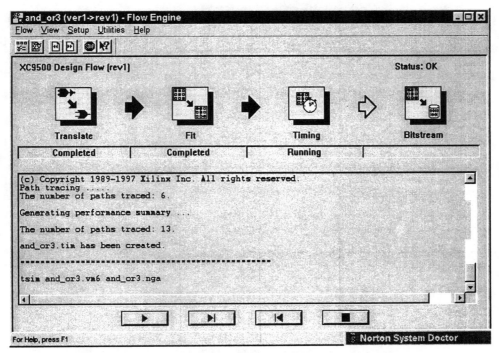

Figure 1.8 Flow Engine window during implementation.

18) Click the **OK** button on the message window.

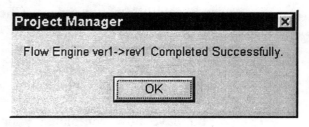

Figure 1.9 Message after step 17.

19) Plug the AC adaptor into an outlet, and plug its cable into your PLDT-1 board. It is important to do that first. Then connect your PC to your PLDT-1 board with the parallel cable.

20) To download the program, click on the **Device Programming** button in the **Project Manager** window (or choose **Tools → Device Programming** on the menu bar).

21) In the **JTAG Programmer** window, choose **Operations → Program** from the menu bar, the **Options** window should occur. In this window, select **Erase Before Programming** and **Verify**, then click on the **OK** button.

22) When the downloading process is finished without error(s), click on the **OK** button in the **Operation Status** window to go back to the **JTAG Programmer** window. Then, choose **File → Exit**. When the message window prompts, select **Yes**. In the **Save As** window, click on the **Yes** button to save the JEDEC file and go back to the **Project Manager** window.

23) Test the program on your PLDT-1 board and fill in the truth tables by using the switches for inputs and LEDs for output. Draw the symbols and write the Boolean equations.

Truth Table for 3-Input AND Gate **Symbol** **Boolean Equation**

A	B	C		X
0	0	0		
0	0	1		
0	1	0		
0	1	1		
1	0	0		
1	0	1		
1	1	0		
1	1	1		

Truth Table for 3-Input OR Gate   **Symbol**   **Boolean Equation**

A	B	C		Y
0	0	0		
0	0	1		
0	1	0		
0	1	1		
1	0	0		
1	0	1		
1	1	0		
1	1	1		

24)   Verify that the experimental results are consistent with our discussion.

Checked by _____   Date _____

NAME _____ DATE _____

QUESTIONS:

1) How many rows must a truth table have in order to describe a 4-input AND gate? Which input state(s) will make the output HIGH?

2) Which output is the unique one in a 3-input OR gate? Does it agree with the statement in our Discussion section?

3) Use a DMM to measure the output voltage levels for logic 0s and logic 1s on the corresponding connectors of T1 and T2 from your PLDT-1 board. Do they match the voltage ranges of logic 0 and logic 1 specified for the CPLD chip? (Refer to Appendix F.)

4) Sketch the output waveform for the given circuit and inputs.

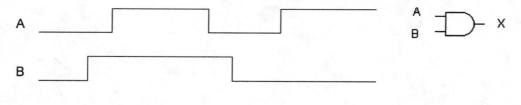

5) In the Xilinx software symbol library, the maximum number of inputs for AND and OR gates is 9. What would you do if 10-input AND and OR gates are needed? Draw the schematic diagrams and show the connections.

CPLD EXPERIMENT 2:
Inverting Logic: NAND, NOR, & NOT

OBJECTIVES:

- Examine inverting logic circuits.
- Demonstrate the characteristics of NOT, NAND, and NOR gates.
- Develop truth tables for NOT, NAND, and NOR gates.

MATERIALS:

- Xilinx Foundation Software, student or professional edition V1.5 or higher.
- IBM or compatible computer with Pentium processor or equivalent, 64 M-byte RAM or more, and 3 G-byte or larger hard drive.
- PLDT-1 board by RSR Electronics Inc., XS95 (V1.2) board by XESS Corp., or a similar board with an XC95108 device.
- Oscilloscope and function generator.

DISCUSSION:

The inverter (or NOT gate) represents logical complementation. A NOT gate can have only one input and one output. The output of a NOT gate simply reverses (inverts) the logic value presented at its input. The NOT gate can be combined with AND and OR gates to construct two more basic gates: NAND and NOR gates. Both NAND and NOR gates are universal logic gates meaning that either NAND gates or NOR gates can be used to construct any combinational logic circuit. We will use gate symbols, truth tables, and Boolean equations to demonstrate their characteristics. As with AND and OR gates, NAND and NOR gates can have two or more inputs but only one output.

Gate Characteristics

1) **The NOT Gate**

Symbol	Boolean Equation	Truth Table

Symbol: A $\rightarrow$ X

Boolean Equation: $X = \overline{A}$

Truth Table

Input Output

A	X
0	1
1	0

Because the NOT gate has only one input, the truth table has two rows. Moreover, the output inverts the logic level of the input. In addition to the overhead bar shown above (read as "X = A-bar"), notation for logical inversion includes the following: !A, /A, ¬A, A*.

2) **The NAND Gate**

Symbol

A
B $\rightarrow$ Y

Boolean Equation: $Y = \overline{A\,B}$

Truth Table

Inputs   Output

A	B	Y
0	0	1
0	1	1
1	0	1
1	1	0

The behavior of a NAND gate can be summarized as follows: The output is LOW only when *all* the inputs are HIGH. If one or more inputs are LOW (false or logic 0), the output will be HIGH. Comparing the truth table for the NAND gate with that of the AND gate, you will find out that each output of a NAND gate is exactly the opposite (inverted) logic value of the corresponding output of an AND gate. In fact, a NAND gate is functionally equivalent to an AND gate cascaded with a NOT gate as shown below.

A
B $\rightarrow$ $\rightarrow$ Y

3)   **The NOR Gate**

| | **Symbol** | **Boolean Equation** | **Truth Table** | | | |

Symbol	Boolean Equation	Truth Table

A
B ▷o— Z $Z = \overline{A + B}$

	Inputs			Output
A	**B**		**Z**
0	0		1
0	1		0
1	0		0
1	1		0

As seen from the above truth table, the output of a NOR gate is HIGH only when *all* the inputs are LOW. If one or more of the inputs are HIGH, then the output is LOW.

Similarly, a NOR gate can be constructed using an OR gate cascaded with a NOT gate. In other words, a NOR gate is functionally equivalent to an OR gate followed by an inverter.

A
B ⊃ ▷o— Z

In the later part of this experiment, we will show how NAND and NOR gates can be used to perform some useful functions such as enabling and disabling signals. Also, we will show how to use NAND and NOR gates to perform the function of a NOT gate.

41

PROCEDURE:

Refer to Tutorial I in APPENDIX C.

1) Start the software by using your mouse to click on the following: **Start** → **Programs** → **Xilinx Foundation Series** → **Xilinx Foundation Project Manager** on your Windows 95/NT/98. You can also start by double clicking the **Xilinx Foundation Project Manager** Icon on your Window.

2) In the **Getting Started** window, choose **Create a New Project** and click on the **OK** button.

3) When the **New Project** window appears, type *INVERT* in the **Name** editor box as your project file name. In the **Directory** edit box, either type or browse to get *Xilabs*. Choose **Type** as *F2.1i*. Click on **Schematic**. Make sure you select the **Family**, **Part**, and **Speed** to be *XC9500*, *95108PC84*, and *20*, respectively. Then, click on the **OK** button.

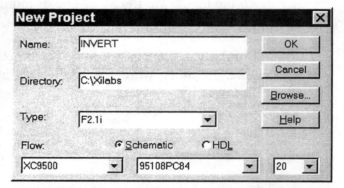

Figure 2.1 New Project window for project *INVERT*.

4) Click on the **Schematic Editor** button in the right pane to get a blank **Schematic Editor** window.

5) Click on the logic symbol icon ⊡. Move the glide bar to INV. The message box indicates: "Inverter". Click your mouse on INV and then click on where you want the symbol to be on the **Schematic Editor** window. The INVERTER symbol should appear. Obtain the NAND3 and NOR3 gate symbols from the SC Symbol window and put them onto the schematic. Your **Schematic Editor** should resemble Figure 2.2 below.

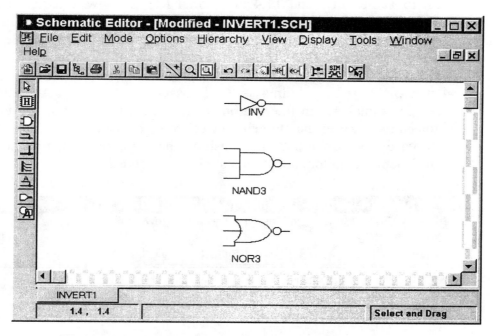

Figure 2.2 Schematic Editor window after step 5.

6) Save this schematic file by using your mouse to choose **File → Save As.** In the **Save As** window, edit the **File name** box to *INVERT.SCH* and click on the **OK** button to save the file to the INVERT subdirectory.

7) Add seven IBUFs and seven input terminals. Name the seven input terminals A, B, C, D, E, F and G from top to bottom. Add three OBUFs and three output terminals. Name the output terminals X, Y, and Z from top to bottom. Then wire all the connections.

8) Refer to Tables T1.3 and T1.4 in Tutorial I to see how to assign the input and output pins. Assign A = pin 6, B = pin 7, C = pin 11, D = pin 5, E = pin 72, F = pin 71, and G = pin 66. This means that we are using toggle switches 1 through 7 on the PLDT-1 board to represent inputs A through G, respectively. Assign LED 1, LED 2, and LED 3 to display the outputs X, Y, and Z. That is, pin 44 = X, pin 43 = Y, and pin 41 = Z. To assign pin 6 to input A, double click the buffer for signal A. In the **Symbol_Properties** window, type *loc* in the **Name** editor box and *p6* in the **Description** editor box. Then, click on the **Add** button followed by the **Move** button. Finish all the pin assignments in the same way. Your Schematic Editor should look similar to Figure 2.3.

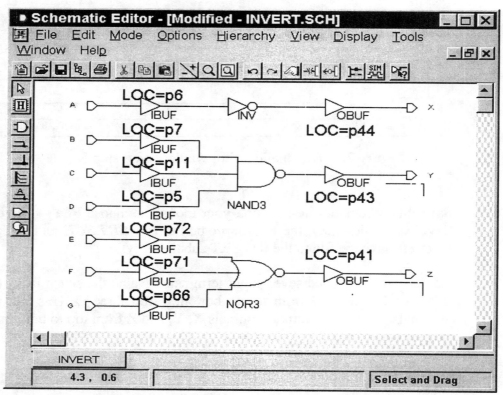

Figure 2.3 Schematic Editor window after step 8.

9) You have finished the schematic drawing for the project INVERT. Save your schematic by choosing **File → Save**.

10) Choose **Option → Create Netlist** on the menu bar. The message should say "Netlist created successfully". Click on the **OK** button.

11) Choose **Option → Integrity Test**. The message "Integrity test passed successfully" should appear. Click on the **OK** button to go back to the **Schematic Editor** window.

12) Select **Option** → **Export Netlist**. Choose *Edit 200 [*.EDN]* for the **Files of type.** Click on the **Open** button to start the exporting process.

13) You have now completed your INVERT design in the **Schematic Editor**. Return to the **Project Manager** window by selecting **File** →**Exit**. Verify that there is a green check mark on the **Design Entry** button.

14) Click on the implementation button to compile your design. Make sure the **Device, Speed, New version name,** and **New revision name** edit boxes have *95108PC84, 20, ver1,* and *rev1,* respectively. You need to stay consistent with the **Device** and **Speed** you chose at the beginning of your schematic design. Click on the **Run** button.

15) Once the implementation process is completed without any errors, you will see the message "Flow Engine ver1→rev1 completed successfully". Click on the **OK** button on the message window and go back to the **Project Manager** window.

16) Plug the AC adaptor into an outlet, and plug its cable into your PLDT-1 board. It is important to do that first. Then connect your PC to your PLDT-1 board with the parallel cable.

17) To program the CPLD chip, click on the **Device Programming** button in the **Project Manager** window (or choose **Tools** → **Device Programming** on the menu bar).

18) In the **JTAG Programmer** window, choose **Operations** → **Program** from the menu bar. The **Options** sub-window should pop up. In this window, select **Erase Before Programming** and **Verify**, then click on the **OK** button.

19) When the downloading process is finished without error(s), click on the **OK** button in the **Operation Status** window to go back to the **JTAG Programmer** window. Then, choose **File** → **Exit**. When the message window prompts, select **Yes**. In the **Save As** window, click on the **Yes** button to save the JEDEC file and go back to the **Project Manager** window.

20) Test the program on the PLDT-1 board and fill in the truth tables by using the switches for inputs and LEDs for output. Draw the symbols and write the Boolean equations.

Truth Table for inverter | Symbol | Boolean Equation

A		X
0		
1		

Truth Table for 3-Input NAND Gate | Symbol | Boolean Equation

B	C	D	Y
0	0	0	
0	0	1	
0	1	0	
0	1	1	
1	0	0	
1	0	1	
1	1	0	
1	1	1	

Truth Table for 3-Input NOR Gate | Symbol | Boolean Equation

E	F	G	Z
0	0	0	
0	0	1	
0	1	0	
0	1	1	
1	0	0	
1	0	1	
1	1	0	
1	1	1	

21) Verify that the experimental results are consistent with the Discussion.

Checked by _____   Date _____

22) Create a Xilinx project called LAB2 in the same way that you did the projects AND_OR3 and INVERT. In this new project prove the following:

a) A 2-input NAND gate is equivalent to a 2-input AND gate followed by a NOT gate.

b) A 2-input NOR gate is equivalent to a 2-input OR gate followed by a NOT gate.

c) A 2-input NAND gate is equivalent to an inverter when the two inputs are tied together.

d) A 2-input NOR gate is equivalent to an inverter when one of the inputs is connected to ground. You can obtain the ground symbol, GND from the SC Symbol window.

You will have 6 inputs and 4 outputs for the schematic. Pick six toggle switches and 4 LEDs on the PLDT-1 board to show your results. Record which switch is for what gate input and which LED is for what gate output since you will be using the results in step 24.

23) Draw the truth tables in the following to demonstrate your results. Do they match what you expected?

a)   b)

c)   d)

24) Sometimes it is necessary to use one or more control signals to either "pass" or "block" one or more input signals. All AND, OR, NAND, and NOR gates can be used to perform this function. When a control signal allows an input signal to pass through a gate, the input signal is said to be *enabled*. In contrast, when a control signal prevents an input signal from passing through a gate, the input signal is said to be *disabled* (or *inhibited*).

Refer to the gate diagrams shown below. Use the design of LAB2 you completed in steps 22 to 24. Set your function generator to output a 10 kHz TTL signal and connect this signal to one input of the NAND gate. Modify your design so that one input of the gate goes to pin 67, the clock input on tie block B2. Connect the function generator to the clock input. Use the input connected to the toggle switch as the control input. Connect channel 1 of your oscilloscope to display the input TTL signal. Use channel 2 of your oscilloscope to display the output. Observe what happens and draw the output waveform corresponding to the input TTL waveform for both a (Control = 0) and b (Control = 1) below. What logic level does the control signal have to take to enable the waveform? What level is required to disable it?

a) **Control Signal B = 0**

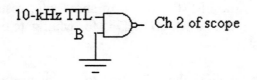

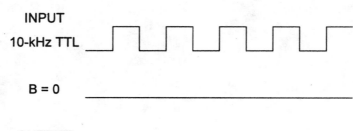

b) **Control Signal B = 1**

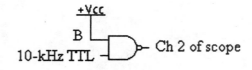

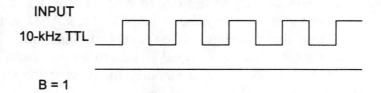

48

QUESTIONS:

1) Show how to construct 3-input NAND and 3-input NOR gates given 2-input NAND gates, 2-input NOR gates and inverters.

2) Can you make an inverter using a 2-input NAND gate in a way other than you showed in procedure 22 c? Draw the circuit.

3) Can you make an inverter using a 2-input NOR gate in a way other than you showed in procedure 22 d? Draw the circuit.

4) Using the same principle as above, show how to make inverters using a 3-input NAND gate and a 3-input NOR gate.

5) In step 24 of the procedure, you proved that the logic level for enabling a NAND gate is HIGH. What logic level should you use to enable a signal to "pass through" a NOR gate?

USE FOR NOTES OR AS A WORK-SHEET

NAME _____ **DATE** _____

CPLD EXPERIMENT 3:
Boolean Laws & Rules and DeMorgan's Theorem

OBJECTIVES:

- Learn and verify Boolean laws and rules.
- Learn and prove DeMorgan's theorem.
- Use Xilinx simulation tools to test combinational circuits.

MATERIALS:

- Xilinx Foundation Software, student or professional edition V1.5 or higher.
- IBM or compatible computer with Pentium processor or equivalent, 64 M-byte RAM or more, and 3 G-byte or larger hard drive.
- PLDT-1 board by RSR Electronics Inc., XS95 (V1.2) board by XESS Corp., or a similar board with an XC95108 device.

DISCUSSION:

A Boolean equation derived directly from a truth table or from a problem statement usually is not in the simplest form. To have an efficient (minimal gates) equivalent logic circuit, the Boolean equation representing the logic design must be in the simplest form. Boolean equations can be simplified using Boolean algebra, DeMorgan's Theorem, or/and Karnaugh maps (K-maps). In this experiment, we will first present Boolean laws and rules as well as DeMorgan's Theorem, and then verify them.

Laws of Boolean Algebra

1) **Commutative Law**

   $A + B = B + A$
   $A\,B = B\,A$ (or, using dot notation, $A \cdot B = B \cdot A$)

2) **Associative Law**

   $A + (B + C) = (A + B) + C$
   $A(B\,C) = (A\,B)C$

3) **Distributive Law**

   $A(B + C) = AB + AC$
   $(A + B)(C + D) = AC + AD + BC + BD$

4) **Operator Precedence**

AND before OR: $A \cdot B + C = (A \cdot B) + C$ ($A \cdot B$ is the same as A B)

These laws are true for any number of variables.

Rules of Boolean Algebra

1) **Operations involving Boolean Constants 0 (FALSE) and 1 (TRUE)**

$$A \cdot 0 = 0 \qquad\qquad A \cdot 1 = A$$
$$A + 0 = A \qquad\qquad A + 1 = 1$$

2) **Operations involving one Boolean Variable and/or its Complement**

Note: The following: A^* , $\overline{A}$, $A\backslash$, $!A$, $\neg A$ are all used to indicate **NOT** A.

$$A \cdot A = A \qquad\qquad A + A = A$$
$$A \cdot A^* = 0 \qquad\qquad A + A^* = 1$$

3) **Use of Logical Equality (=)**

Note: AB is the same as $A \cdot B$

$$A^{**} = A$$
$$A + A^* B = A + B \qquad (\text{the absorption rule})$$
$$A^* + AB = A^* + B$$

We can use the rules of Boolean algebra to combine or eliminate certain variables in the Boolean equations to obtain simpler equivalent circuits.

DeMorgan's Theorem

$$\overline{A + B} = \overline{A}\ \overline{B} \qquad\qquad\qquad \overline{A\ B} = \overline{A} + \overline{B}$$

Notice that DeMorgan's theorem is a pair of Boolean equations. The first one in the pair states that a NOR gate is equivalent to an AND gate with inverted inputs. The second one reveals that a NAND gate is equivalent to an OR gate with inverted inputs. Moreover, DeMorgan's Theorem applies to two or more variables.

PROCEDURE:

Refer to Tutorials I and II in Appendices C and D.

Section I. Boolean Distributive Law

1) Start the software by clicking on the following: **Start → Programs → Xilinx Foundation Series → Xilinx Foundation Project Manager** on your Windows 95/NT/98. You can also start by double clicking the **Xilinx Foundation Project Manager** Icon on your Windows desktop.

2) In the **Getting Started** window, choose **Create a New Project** and click on the **OK** button.

3) When the **New Project** window appears, type *B_LAWS* in the **Name** editor box as your project file name. In the **Directory** editor box, either type or browse to get *Xilabs*. Choose **Type** as *F2.1i*. Click on **Schematic**. Make sure you select the **Family**, **Part**, and **Speed** to be *XC9500*, *95108PC84*, and *20*, respectively. Then, click on the **OK** button.

4) Click on the **Schematic Editor** button to get a blank **Schematic Editor** window.

5) Since there are in total three Boolean laws and each has two forms as listed in the previous section of this lab, we will verify only the last one:

$$(A + B)(C + D) = AC + AD + BC + BD$$

The rest of the Boolean laws can be verified in the same way.

We can name X1 and X2 as the outputs of the two logic circuits represented by the two sides of the Boolean equation. That means we break the above equation into two:

$$X1 = (A + B)(C + D)$$
$$X2 = AC + AD + BC + BD$$

If we can show that X1 and X2 are equal for all the input combinations, we will have proved this Boolean law. (When done with truth tables, this technique is called "Proof by Perfect Induction".)

On the **Schematic Editor** window, build the logic circuit similar to the one shown in Figure 3.1 below.

Notice that the two circuits are combined into one since they share the same set of inputs: A, B, C and D. Observe the schematic circuit in Figure 3.1. The top part represents X1 = (A + B)(C + D) and the bottom part stands for X2 = AC + AD + BC + BD. The two 2-input OR gates give us the two sum terms A+B and C+D, and then they are ANDed by a 2-input AND gate to form the X1 equation. The four product terms AC, AD, BC and BD are realized by four 2-input AND gates, and then they are ORed using a 4-input OR gate to provide us with X2.

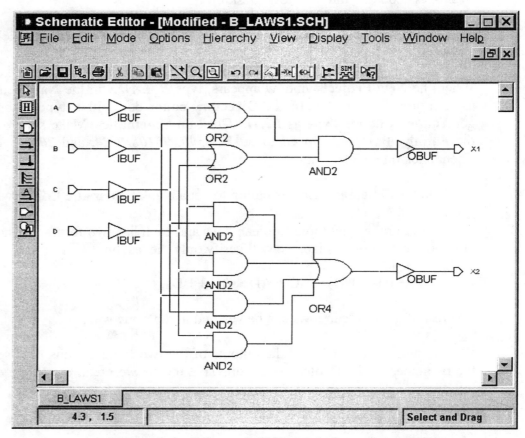

Figure 3.1   Schematic Editor window after step 5.

6)   Save this schematic file by choosing **File → Save As**. In the **Save As** window, edit the **File name** box to *B_LAWS.SCH* and click on the **OK** button to save the file to the B_LAWS subdirectory.

7) We need to use four switches and two LEDs to represent the four inputs and two outputs. Assign A = pin 6, B = pin 7, C = pin 11, and D = pin 5. This means that we will use toggle switches 1 through 4 on the PLDT-1 board to represent inputs A through D, respectively. Assign LED 1 and LED 2 to display the outputs X1 and X2. That is, pin 44 = X1 and pin 43 = X2. If you need to review how to assign input and output pins, refer to Lab 2. Once you have finished the pin assignment, your **Schematic Editor** window should look similar to Figure 3.2.

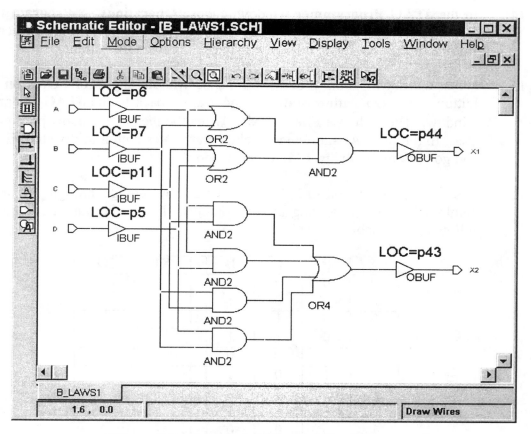

Figure 3.2 Schematic Editor window after step 7.

8) Save your schematic again by choosing **File → Save**.

9) Go through **Option → Create Netlist, Option → Integrity Test,** and then **Option → Export Netlist** as specified for the previous two labs. Return to the **Project Manager** window by selecting **File → Exit**.

10) Click on the implementation button to compile your design. Make sure the **Device, Speed, New version name,** and **New revision name** edit boxes have *95108PC84, 20, ver1,* and *rev1,* respectively.

11) When the implementation process is completed without any errors, click on the **OK** button on the message window and go back to the **Project Manager** window.

12) Plug the AC adaptor into an outlet, and plug its cable into your PLDT-1 board. Then connect your PC to your PLDT-1 board with the parallel cable.

13) Program your CPLD chip by clicking on the **Device Programming** button in the **Project Manager** window (or choose **Tools** → **Device Programming** on the menu bar).

14) In the **JTAG Programmer** window, choose **Operations** → **Program** from the menu bar, the **Options** window should occur. In this sub-window, select **Erase Before Programming** and **Verify**, then click on the **OK** button.

15) When the downloading process is finished without any errors, click on the **OK** button in the **Operation Status** window to go back to the **JTAG Programmer** window. Then, choose **File** → **Exit**. When the message window prompts, select **Yes**. In the **Save As** sub-window, click on the **Yes** button to save the JEDEC file and go back to the **Project Manager** window.

16) Test the program on the PLDT-1 board by going through all the input combinations and observing the two outputs. Fill in the output columns in the following truth table.

A	B	C	D		X1	X2
0	0	0	0			
0	0	0	1			
0	0	1	0			
0	0	1	1			
0	1	0	0			
0	1	0	1			
0	1	1	0			
0	1	1	1			
1	0	0	0			
1	0	0	1			
1	0	1	0			
1	0	1	1			
1	1	0	0			
1	1	0	1			
1	1	1	0			
1	1	1	1			

Table 3.1 Test Data

17)   Are the two output columns the same? If they are, you have proved the Boolean distributive law. If not, you may want to check:

a)   Are all the logic gates what they are supposed to be?
b)   Do the inputs connected to the gates give you the correct sums and products?
c)   Are you using the assigned switches (1 to 4) and LEDs (1 and 2)?

Once you find the problem, repeat the implementation and programming steps. Test your circuit again.

Checked by _____ Date _____

You can also use the Xilinx simulation tools to verify the Boolean distributive law. Refer to Tutorial II for the details of how to perform a simulation for a combinational logic circuit.   Assume that you have completed the previous 17 steps and you are in the **Project Manager** window, which has the *B_LAWS* project open.

18)   Click on the **Simulation** button to get into the **Logic Simulator** window. Choose **Signal → Add Signals** from the menu bar. Add all the input signals (A, B, C, and D) and then the output signals (X1 and X2) in the order shown on Figure 3.3. From Tutorial II, adding a signal is done by selecting the signal name and then clicking on the **Add** button in the **Component Selection for Waveform Viewer** window. When you finish adding all the signals, close the **Component Selection for Waveform Viewer** window to go back to the **Logic Simulator** window.

19)   In the **Logic Simulator** window, choose **Signal → Add Stimulators** from the menu bar. In the **Stimulator Selection** window, add B0, B1, B2 and B3 to A, B, C and D, respectively. To do this, select the signal in the **Logic Simulator** window first, and then click on the stimulator button. When you finish the stimulator selections, click on the **Close** button to go back to the **Logic Simulator** window.

20)   To start the simulation, choose **Options → Start Long Simulation**. Click on the **On** button on the menu bar to shift the waveform to the beginning. Click on the **Start Simulation** button ⬛. To view the waveforms clearly, you may have to scale the waveform by clicking on the **Zoom In** ▥ or **Zoom Out** ▥ buttons. When you finish these, your **Logic Simulator** window should look similar to Figure 3.3. Notice that the two output waveforms for X1 and X2 look exactly the same for all the combinations of the inputs. More than two cycles of the waveforms are displayed in the **Waveform Viewer** window.

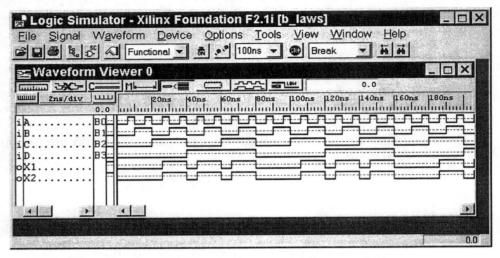

Figure 3.3 Simulation result for Boolean Distributive law.

Checked by _____ Date _____

Section II. Boolean Absorption Rule

Observe the list of Boolean Rules in the discussion section of this experiment; you will see 11 Boolean equations. The two shown below are called absorption rules.

$$A + \overline{A}B = A + B$$

$$\overline{A} + AB = \overline{A} + B$$

You are asked to prove them using simulation tools.

1) In the **Project Manager** window, open a new project called ***B_RULES***.

2) In the **Schematic Editor** window, build one circuit to verify the two absorption equations. You can use two inputs (A and B) and four outputs (X1, X2, Y1, and Y2). In other words, you can break the above two equations into four:

$$X1 = A + \overline{A}B \qquad\qquad X2 = A + B$$

$$Y1 = \overline{A} + AB \qquad\qquad Y2 = \overline{A} + B$$

If you show that the output waveforms X1 and X2 are the same and that the output waveforms Y1 and Y2 are the same for all the input combinations, you have verified the Boolean absorption rule.

3)   Once you finish the schematic, you can go to the simulation directly without going through the implementation and programming steps. Since you will not program your CPLD for this design, you will not need to make any pin assignments. Demonstrate your final **Waveform Viewer** window.

4)   Fill the following table based on the waveforms you obtained from step 3.

A	B		$X1 = A + \bar{A}B$	$X2 = A + B$	$Y1 = \bar{A} + AB$	$Y2 = \bar{A} + B$
0	0					
0	1					
1	0					
1	1					

Table 3.2 Data from Waveforms

Checked by _____ Date _____

Section III. DeMorgan's Theorem

1)   Create a project called **_DEMORGAN_**. Build a circuit that enables you to prove the two equations of DeMorgan's theorem. You will need two inputs (A and B) and two pairs of outputs (X1, X2, Y1 and Y2). Similar to section II of this experiment, one output is for the first equation and another is for the second equation in DeMorgan's theorem.

2)   Use the simulation tool to generate the input and output waveforms. Draw the waveforms in the space provided below:

A

B

X1

X2

Y1

Y2

59

3) Fill the following truth table based on your simulation result.

A	B		$X1 = \overline{A + B}$	$X2 = \overline{A}\,\overline{B}$	$Y1 = \overline{A\,B}$	$Y2 = \overline{A} + \overline{B}$
0	0					
0	1					
1	0					
1	1					

Table 3.3 Data from Simulation Results

4) Comment on your results in this space:

Checked by _____ Date _____

QUESTIONS:

1) Draw the logic diagrams and list the truth tables for all the Boolean algebra rules except the absorption rule.

2) Apply Boolean laws and rules and DeMorgan's theorem to simplify the following Boolean equations. Draw the simplified logic diagrams:

a) $X = (A + B)\overline{ABC} + \overline{BC}$

b) $Y = (\overline{A + B})\overline{B} + B + \overline{AC}$

NAME _____ **DATE** _____

CPLD EXPERIMENT 4:
XOR and XNOR Gates with Applications

OBJECTIVES:

- Examine the characteristics of Exclusive-OR (XOR) and Exclusive-NOR (XNOR) gates.
- Demonstrate applications of XOR and XNOR gates.

MATERIALS:

- Xilinx Foundation Software, student or professional edition V1.5 or higher.
- IBM or compatible computer with Pentium processor or equivalent, 64 M-byte RAM or more, and 3 G-byte or larger hard drive.
- PLDT-1 board by RSR Electronics Inc., XS95 board and XStend board by XESS Corp., or a similar board with an XC95108 device.
- Oscilloscope and function generator.

DISCUSSION:

So far we have studied five basic types of gates: AND, OR, NAND, NOR and NOT. In some applications, it is convenient to use two other types of gates: XOR and XNOR. The XOR and XNOR gates have their own symbols and unique characteristics. Common applications for XOR and XNOR gates are: comparators, switchable inverter/buffers, parity generator/checkers, and adder/subtractor circuits. They can also be used to simplify Boolean equations.

We will first discuss the properties of XOR and XNOR gates having two inputs.

Gate Characteristics

1) The XOR Gate

Symbol	Boolean Equation	Truth Table

$$X = \overline{A}B + A\overline{B}$$

Inputs		Output
A	**B**	**X**
0	0	0
0	1	1
1	0	1
1	1	0

For a 2-input XOR gate, the output is HIGH when the inputs are unequal. The output is LOW when the inputs are equal. The Boolean equation for a 2-input XOR gate can be briefed as: $X = A \oplus B$. However, the function definition remains the same.

2) The XNOR Gate

Symbol	Boolean Equation	Truth Table

$$Y = AB + \overline{A}\,\overline{B}$$

Inputs		Output
A	**B**	**Y**
0	0	1
0	1	0
1	0	0
1	1	1

The output of a XNOR gate is the complement of that of a XOR. For a 2-input XNOR gate, the output is LOW when the inputs are unequal but HIGH when the inputs are equal. The Boolean equation for a 2-input XNOR gate can be written as:

$$Y = \overline{A \oplus B}$$

The number of inputs for the XOR and XNOR gates can be two or more. The characteristics of XOR and XNOR gates can be extended to three or more inputs. We will examine the characteristics of 3-input XOR and XNOR gates below.

PROCEDURE:

Section I.   XOR and XNOR Characteristics

1)   Start a new project called EXGATES.

2)   In the **Schematic Editor** window, include a 3-input XOR (XOR3) and a 3-input XNOR (XNOR3). You will have three inputs: A, B, and C. You will need two outputs, X and Y, representing the outputs of the XOR3 and XNOR3 gates respectively.

3)   When you finish the schematic design, use the simulation tool to obtain the complete results for both XOR3 and XNOR3 gates. Based on the waveforms, fill in the following truth tables.

A	B	C		Output of XOR3 X	Output of XNOR3 Y
0	0	0			
0	0	1			
0	1	0			
0	1	1			
1	0	0			
1	0	1			
1	1	0			
1	1	1			

Table 4.1 Truth Table for 3-Input XOR and 3-Input XNOR Gates

4)   The above truth table should reveal:

a)   For an XOR gate, the output is LOW when the number of 1 bits in the input is an EVEN number and the output is HIGH when the number of 1 bits in the input is an ODD number.

b)   The output of an XNOR gate is the complement of that of a XOR gate.

These results can be extended to XOR and XNOR with more than three inputs.

Checked by _____   Date _____

Section II.   XOR Gates Used In A Comparator

A simple 4-bit comparator will be built in this section of the experiment.   This comparator does not distinguish which 4-bit binary word is larger or smaller.   It only informs us whether the two words are the same or not.   Specifically, the output of this comparator is HIGH when the two 4-bit binary words are the same but LOW when the two words are different.   Although the comparator is implemented with XOR gates here, you can implement the same function using XNOR gates coupled with a 4-input AND gate.

1)   Start a new project called **COMPARA**.

2)   Build a schematic design similar to Figure 4.1 below.   In this design, we will use all eight switches (1 to 8) to represent the two binary words 'A' (A0, A1, A2, A3) and 'B' (B0, B1, B2, B3) from top to bottom, and LED 1 to display the output.

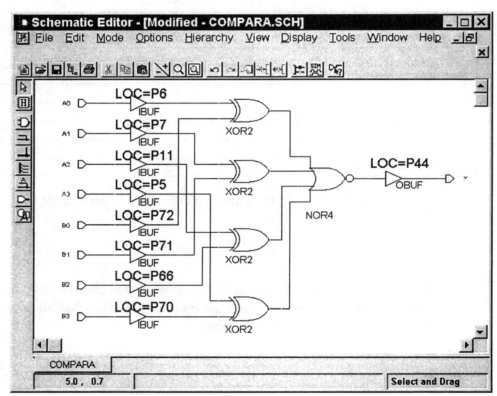

Figure 4.1 Schematic design for the 4-bit comparator.

3)   Save this schematic design and return to the **Project Manager** window and implement this design.

4)   When you have completed the implementation successfully, download the JEDEC program to your target board.

5) Test your circuit on the target board by using the switches to enter all 1/0 combinations of the inputs as shown below. Fill in the truth table and comment on the result.

Word A					Word B				Output
A0	A1	A2	A3		B0	B1	B2	B3	X
SW 1	SW 2	SW 3	SW 4		SW 5	SW 6	SW 7	SW 8	LED 1
0	1	1	0		1	0	0	1	
1	0	1	0		1	0	1	0	
0	0	1	1		1	1	0	0	
0	1	0	0		0	1	0	1	
1	1	0	1		1	1	0	1	
0	0	0	0		0	0	0	0	
0	1	1	1		0	1	1	1	
1	1	1	1		1	1	1	0	
1	0	1	1		0	0	1	1	

Table 4.2 Results for Comparing Words A and B

Checked by _____ Date _____

67

Section III. XNOR Gates Used As Buffers and Inverters

If you examine the truth table of an XNOR gate (refer to Part I) carefully, you will notice an interesting fact: When input A is held LOW, the output is the complement of input B. When input A is kept HIGH, the output follows input B. This effect means that the XNOR gate can be used to construct a buffer/inverter circuit. What we have to do is use one of the inputs as the control signal and the other input as the data signal. The XNOR will act like a buffer when the control signal is HIGH, but as an inverter when the control signal is pulled LOW. Here we will build a 4-bit buffer/inverter circuit and then run a simulation to verify the result.

1) Create a new project called BUF_INV.

2) Build a schematic design similar to Figure 4.2 below. We will need five switches. Switches 1 to 4 are used to supply four data bits: D0, D1, D2, and D3, respectively. Switch 8 is used to input the control signal (CONTROL). Four LEDs (1 to 4) are used to display the 4-bit output data X0, X1, X2, and X3.

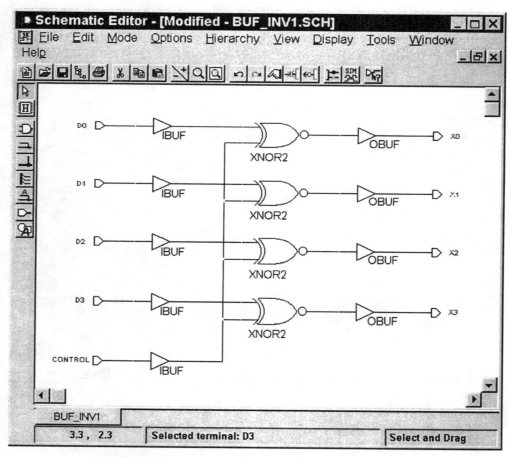

Figure 4.2 A 4-bit buffer/inverter circuit.

3)   Use the simulation tool to obtain waveforms similar to Figures 4.3 (a) and (b) below. Notice that the four data inputs D0, D1, D2, and D3 are connected to the counter stimuli B0, B1, B2, and B3 to generate all possible 1/0 combinations on the inputs. The control input is connected to a toggle switch (key 'q') in the **Keyboard** area of the **Simulation Selection** window.

To simulate CONTROL = 0, just click on the line next to '**q**' in the **Waveform Viewer** window to make this input LOW. Then choose **Options → Start Long Simulation…** from the menu. If you want align the waveform to start from t = 0, click the "ON" button. Now you should have a set of waveforms as shown in Figure 4.3 (a).

To simulate CONTROL = 1, click the line next to '**q**' to make this input HIGH. Again, choose **Options → Start Long Simulation…** from the menu. You should get a set of waveforms similar to the ones in Figure 4.3 (b).

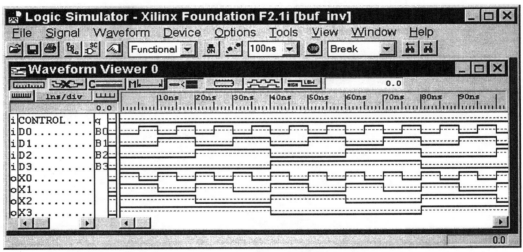

Figure 4.3 (a) Simulation result for CONTROL = 0.

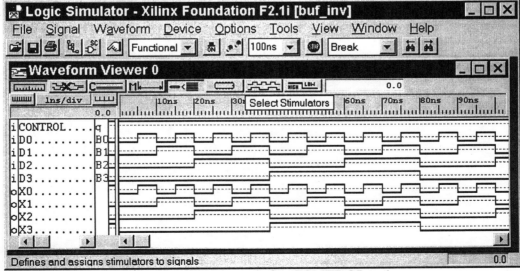

Figure 4.3 (b) Simulation result for CONTROL = 1.

69

4) Compare the waveforms D0 through D1 with the waveforms X0 through X2 in Figure 4.3 (a) above. Is the circuit now a buffer or an inverter?

5) Compare the waveforms D0 through D1 with the waveforms X0 through X2 in Figure 4.3 (b) above. Is the circuit now a buffer or an inverter?

Checked by _____ Date _____

QUESTIONS:

1)   A 3-input XOR gate is equivalent to the circuit shown below:

The Boolean equation can be written as:

$$X = (\overline{\overline{A}\,B + A\,\overline{B}})C + (\overline{A}\,B + A\,\overline{B})\,\overline{C}$$

Or simply denoted as:

$$X = A \oplus B \oplus C$$

Using only AND, OR, and inverter gates to implement the above Boolean equation, how many gates are needed? Draw the logic diagram. Compare the savings of an single XOR gate implementation with the circuit you just drew.

2)   How can you use a 2-input XOR gate to function as a 1-bit buffer/inverter? Draw the logic diagram. Show the logic connections for the control and data input lines.

3)   Draw the logic circuit for each of the following. For each gate, determine EVEN or ODD parity and find the output for the given input data:

a)   4-input XOR, input data = 1001

b)   5-input XNOR, input data = 10010

c)   6-input XOR, input data = 101001

d)   7-input XNOR, input data = 10110110

CPLD EXPERIMENT 5:
Use of XOR / XNOR Gates to Generate & Check Parity

OBJECTIVES:

- Demonstrate the use of XOR and XNOR gates to generate and check parity.

MATERIALS:

- Xilinx Foundation Software, student or professional edition V1.5 or higher.
- IBM or compatible computer with Pentium processor or equivalent, 64 M-byte RAM or more, and 3 G-byte or larger hard drive.
- PLDT-1 board by RSR Electronics Inc., XS95 board and XStend Board by XESS Corp., or a similar board with an XC95108 device.

DISCUSSION:

A parity checker circuit is used to detect a 1-bit error, as could occur in data transmission. Such an error can be caused by an electrical noise "hit", or by a hardware failure such as a bit "stuck-at-0" or "stuck-at-1". Parity can be even or odd and requires that an extra bit (the parity bit) be generated and "tacked onto" the data. The value of the parity bit is derived from the data. The following examples illustrate even and odd parity.

1)   **Even Parity**

Determine the parity bit for the 4-bit data 1001 using even parity.

The number of 1 bits in 1001 is 2, an even number. To have even parity for this data, we must keep the total number of 1s, including the parity bit, even. Hence, the parity bit should be a 0. Note that, as a binary number, 1001 is odd (value is 9). But whether the value of 1001 is odd or even is not important. Only the number of 1s is important.

Data: 1001   Parity Bit: 0   Data with Parity: 10010

2)   **Odd Parity**

Determine the parity bit for the 8-bit data 10110111.

The number of 1 bits in 10110111 is 6, an even number. To have odd parity for this data, we need to make sure the total number of 1s, including the parity bit, is an odd number. Therefore, the parity bit must be 1. Again, the numerical value of the data is not important. All that counts is the number of 1s in the data.

Data: 10110111   Parity Bit: 1   Data with Parity: 101101111

PROCEDURE:

We will build an even parity generator/checker for an 8-bit transmitter/receiver. The output ERROR should be HIGH when there is a 1-bit error during transmission and LOW when there is no error. (While errors in data communications typically are "bursty", for this example we will assume that only 1-bit errors can occur.) We will use an 8-bit XOR gate for the parity generator (in the transmitter section) and a 9-bit XOR gate for the parity checker (in the receiver section). Since we have 8 data bits, it's convenient to use the symbol IBUF8, which is a group of eight IBUFs. Let's start building this project.

1)   Create a new project called **GEN_CHK**.

2)   Retrieve an XOR8 gate, an XOR9 gate, two IBUF8s and two OBUFs from the **SC Symbol** window and place them in the proper locations on your **Schematic Editor** window as shown in Figure 5.1 below.

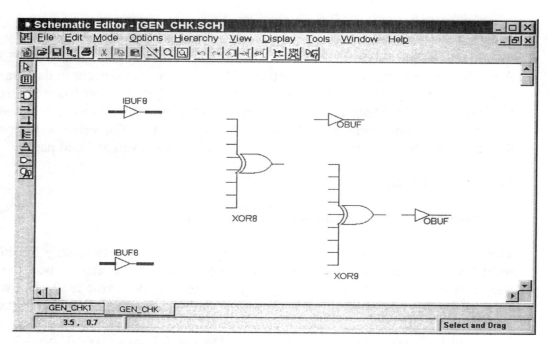

Figure 5.1 Schematic Editor window for project **GEN_CHK**.

3)   To draw the solid bus lines that connect the output of IBUF8 to the inputs of the XOR8 gate, click on the **Draw Buses** button (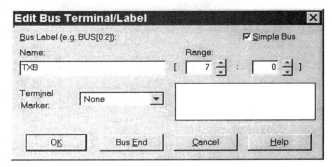) or select **Mode → Draw Buses** from the menu on the **Schematic Editor** window. Start the solid line at the output of IBUF8 and pull it downward to the point where you want it to end. To end the bus line, double click to get the **Add Bus Terminal/Label** window. Since you only want to draw the bus line, not the terminals, select *None* in the

Terminal Marker selection box. Type *TXB* (transmitter bus) in the **Name** editor box. Choose the **Range** from *7* to *0*. Then, click on the **OK** button to go back to your **Schematic Editor** window. You may move the bus name by selecting and dragging it to the desired location.

Figure 5.2 Add Bus Terminal/Label window.

4)   To obtain the input terminals for the *TXB* bus, start the bus line on the input of IBUF8 and drag the solid line to the left to where you want to place the terminals. Double click to get the **Add Bus Terminal/Label** sub-window, then choose *Input* for the **Terminal Marker** selection box since you are making input terminals. Type *TX* (transmitting signal) in the **Name** editor box and select the **Range** *7* to *0*. Then, click on the **OK** button. Repeat the same process for the bus lines *RXB* (receiving bus) and input terminals *RX* (receiving signal).

5)   Add two output terminals, *EP* (even parity) and *ERROR*.

6)   To make the connections from XOR8 and XOR9 to the two buses, click on the **Draw Bus Taps** button (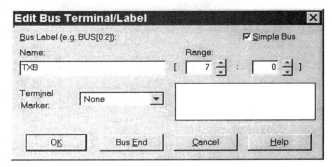) or select **Mode → Draw Bus Taps** from the menu on the **Schematic Editor** window. Then, click on the top input to the XOR8 gate. The connection should appear. Repeat the same for all other wires to the buses. Make other connections as necessary.

7)   Label all the wires that represent transmitted data lines and received data lines. To do this, double click on the top wire connecting to XOR8. In the **Net Name** sub-window, type *TXB0*. Then, click on the **OK** button. Repeat this process for all other wires (*TXB1* to *TXB7*, *RXB0* to *RXB7*) connecting to the *TXB* and *RXB* buses.

Figure 5.3 Net Name sub-window

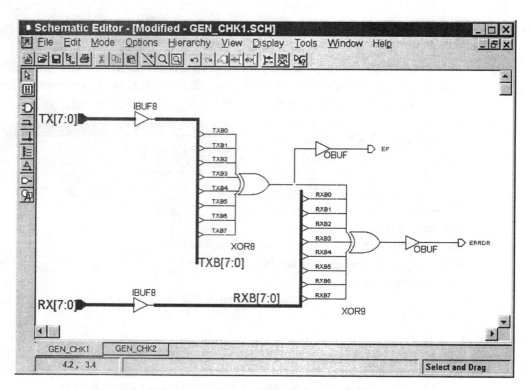

Figure 5.4 Schematic Editor window after step 7.

8) Your schematic should now look like Figure 5.4 above. Save the schematic. Then go through **Option → Create Netlist, Options → Integrity Test,** and **Options → Export Netlist ...** menu items to create, check, and export the netlist for the 8-bit even parity generator/checker. Then close the **Schematic Editor** window.

9)   We can now perform a simulation for this design. Click on the **Simulation** button on your **Schematic Editor** window. Choose **Signal → Add Signals...** to add *TX*, *RX*, *EP,* and *ERROR* signals, respectively. Your **Logic Simulator** window should look similar to Figure 5.5 below.

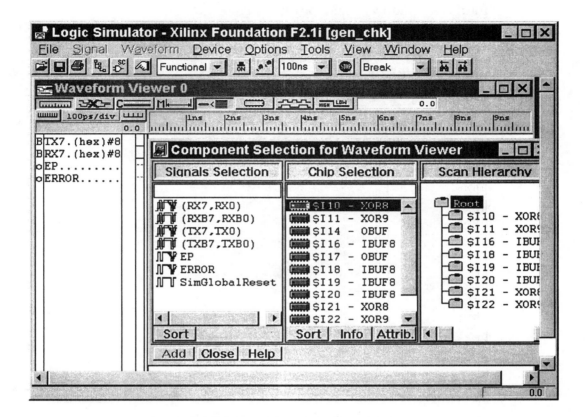

Figure 5.5   Logic Simulator window after step 9.

10)   Since there are 8 different lines in the *TX* bus and 8 different lines in the *RX* bus, we must expand them into individual lines in order to assign each one a stimulator. To do this, click on the *TX* bus name in the **Waveform Viewer** window. Then choose **Signal** → **Bus** → **Flatten** from the menu on the **Logic Simulator** window. Repeat the same process for the *RX* bus. Then, close the **Component Selection for Waveform Viewer** sub-window. Your **Logic Simulator** window should look like Figure 5.6 below.

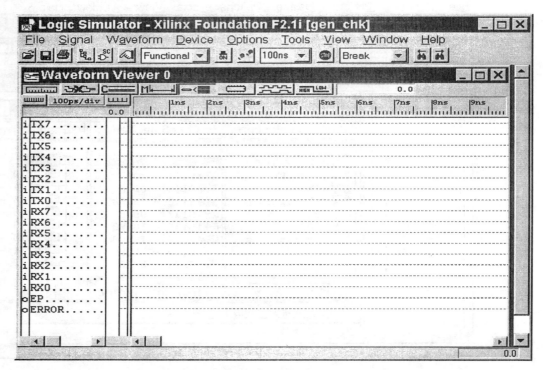

Figure 5.6   Logic Simulator window after step 10.

11)   To test if this schematic design works, we need to assign logic values to all the transmitted data lines and received data lines. According to our design, if the received data are the same as the transmitted data, then the output *ERROR* = 0. If there is one bit different between transmitted data and received data (i.e., a 1-bit error occurred), then *ERROR* = 1. The logic value of *EP* will be determined by the specific data assigned to the *TX* input.

We will not use a binary counter to generate all 256 possible input combinations. Rather, we will assign one set of bit values to the inputs and look at two cases for the output in steps 12 and 13 below.

12)   Case I: TX[7:0] = 10110001 and   RX[7:0] = 10110001

The received data is the same as the transmitted data. So we expect to see:

EP = 0   the input has even number of 1s
ERROR = 0   no error occurs

Open the **Stimulator Selection** window by selecting **Signal → Add Stimulators**. Find the buttons for "1" and "0" in the **Keyboard** area of the **Stimulator Selection** window.   Click on the name *TX7* in the **Waveform Viewer** window and then click on the "1" button in the **Keyboard** area of the **Stimulator Selection** window.   A "1' appears in the column next to *TX7*. Repeat this process to assign logic values to *TX6* through *TX0* (0110001), and *RX7* through *RX0* (10110001).   Your **Logic Simulator** window should now look like Figure 5.7 below.

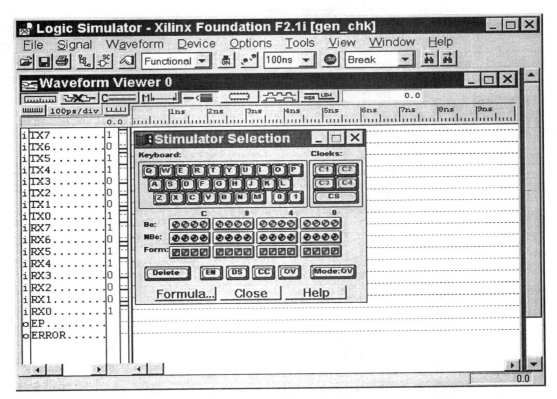

Figure 5.7   Logic Simulator window after step 12.

Close the **Stimulator Selection** window. Choose **Option** → **Start Long Simulation...**, and then click on the "ON" button. The **Waveform Viewer** window should look like Figure 5.8 below. Observe that the outputs *EP* and *ERROR* are both LOW, as we expected.

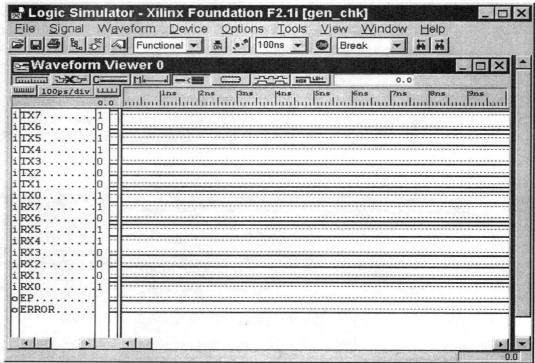

Figure 5.8 Logic Simulator window after step 13.

13)   Case II: TX[7:0] = 10110001 and RX[7:0] = 10110101

TX[7:0] remains the same. One bit of RX[7:0], namely RX2, is stuck at 1. All other RX bits are OK. Then we expect to see:

EP = 0   no change on TX and the number of 1s is even
ERROR = 1   1-bit error occurs

Change the stimulator for RX2 from 0 to 1. After we have assigned the logic values for all the inputs, we can "bundle" the eight lines in a group and view them as a hexadecimal number in the **Waveform Viewer** window. That can make the observation easier. To combine the eight TX signals, first click on TX7, then hold the **Shift** key while you click TX0. You should see TX7 through TX0 are highlighted. Then, select **Signal** → **Bus** → **Combine**. TX7 through TX0 are combined now. Repeat this process for RX7 through RX0.

80

To initiate the simulation, choose **Option → Start Long Simulation...** and click on the "ON" button. Your **Logic Simulator** window should look like Figure 5.9 below. It is shown that EP = 0 since we did not change TX[7:0] and the number of 1's in TX[7:0] is even. The output ERROR = 1, indicating a 1-bit error has occurred, caused by TX2 being stuck at 1 where the correct logic value should be 0. Note that hexadecimal 8D is the equivalent to binary 10001011, which assumes that TX0 is the most significant bit.

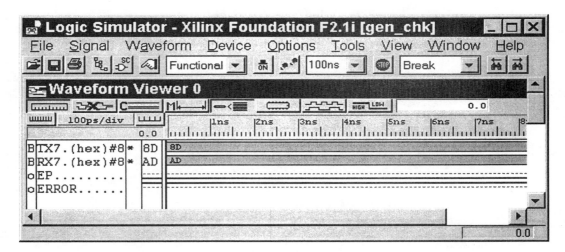

Figure 5.9 Logic Simulator window after step 14.

14) Repeat the simulation process using the four sets of transmitted and received bits given in the following table. In each case, find the logic values for the outputs EP and ERROR. Note that you need to flatten the combined TX and RX before assigning logic values to them.

TX7 TX0	RX7 RX0	EP	ERROR
1 1 1 0 1 1 1 1	1 1 1 0 1 1 1 1		
0 1 1 1 0 1 0 0	0 1 1 1 0 1 0 1		
0 0 0 0 0 1 0 0	0 0 0 0 0 1 0 0		
1 1 0 0 0 0 0 1	1 0 0 0 0 0 0 1		

15) In this space, explain the results you obtained above.

Checked by _____ Date _____

81

NAME _____ DATE _____

QUESTIONS:

1)   We discussed how to construct an 8-bit even parity generator/checker using XOR gates. Can you use XNOR gates to perform the same function? Draw the schematic diagram.

2)   If you are asked to build an 8-bit odd parity generator/checker with the output ERROR = 1 when there is a 1-bit error, which type of gates, XOR or XNOR is simpler to use? Explain.

3)   Why can't a parity generator/checker detect errors of more than 1-bit? Why can it not detect, say, 2-bit errors? Give an example to explain.

CPLD EXPERIMENT 6:

Karnaugh Maps

OBJECTIVES:

- Use Karnaugh maps (K-maps) to simplify Boolean equations.
- Examine the relationship between a K-map and the truth table it represented.

MATERIALS:

- Xilinx Foundation Software, student or professional edition V1.5 or higher.
- IBM or compatible computer with Pentium processor or equivalent, 64 M-byte RAM or more, and 3 G-byte or larger hard drive.
- PLDT-1 board by RSR Electronics Inc., XS95 board and XStend Board by XESS Corp., or a similar board with an XC95108 device.

DISCUSSION:

We used Boolean algebra to simplify logic equations in Experiment 3. Karnaugh mapping is an alternate way to simplify such equations. The K-map is a graphical method based on the laws and rules of Boolean algebra. In practice, K-maps are used to simplify Boolean equations with six or fewer inputs only. Larger K-maps are too unwieldy to be practical.

Valid K-Map Layouts

Essentially, a K-map consists of a box subdivided into rows and columns forming squares called cells. Each row and column will be associated with one or more of the variables. The number of cells equals 2 to the power N, where N is the number of the input variables. For example, if the number of inputs is 4, there are 16 squares. Moreover, the squares must be grouped so that the number of rows and the number of columns are either the same or differ by one. For example, if there are three input variables, you would have two rows and one column, or vice versa. If there are four input variables, then there are two rows and two columns.

The input variables are used to label the rows and columns in a K-map. The variables are assigned to the rows and columns such that only one variable changes between any two adjacent rows or any two adjacent columns. Now here is a crucial property of K-maps: the upper edge and the lower edge of the map are the same edge! Likewise, the left edge and the right edge of the map are the same edge. To see how that is possible, try to visualize the K-map as drawn on the surface of a doughnut (what mathematicians call a "torus"). So a cell along the top edge is adjacent to a cell along the bottom edge and a cell along the left edge is adjacent to a cell along the right edge.

Some valid K-map layouts are illustrated in the following.

a)   2-input variable K-map

	$\overline{B}$	B
$\overline{A}$		
A		

b)   3-input variable K-map

	$\overline{C}$	C
$\overline{A}\,\overline{B}$		
$\overline{A}\,B$		
$A\,B$		
$A\,\overline{B}$		

or

	$\overline{B}\,\overline{C}$	$\overline{B}C$	BC	$B\overline{C}$
$\overline{A}$				
A				

c)   4-input variable K-map

	$\overline{C}\,\overline{D}$	$\overline{C}D$	CD	$C\overline{D}$
$\overline{A}\,\overline{B}$				
$\overline{A}\,B$				
$A\,B$				
$A\,\overline{B}$				

84

Karnaugh Mapping Rules

1) Before mapping a Boolean equation, the equation has to be converted into either a sum-of-products (SOP) or a product-of-sums (POS) form if it is not already in such form. The SOP expression may be easier to use than the POS expression, or vice versa, depending on the specific problem. Since the mappings for the two forms are similar, we will only discuss K-maps for the sum-of-products form.

2) After obtaining the Boolean equation in the SOP form, draw the K-map layout. While not strictly necessary, it is easier if we systematically assign row and column variables from left to right. For example, if the input variables are A, B, C, and D, use A and B for row variables, and C and D for column variables.

3) A "full term" is one that contains all the variables. A "partial term" is one that does not contain all the variables. For each full product term in the Boolean equation, a 1 is put into a corresponding square in the K-map. Identify the square by finding the intersection between the row and the column represented by the variables. For a partial term, 1s are put into two or more cells. For example, if A, B, C, and D are the variables and AB is a product term, $2^2 = 4$ squares should be filled with 1s. The multiple 1s represent "don't cares", meaning the term is true whether the missing variables (C & D) are true or false.

4) Encircle groups where all the adjacent squares contain 1s, but only groups that are powers of 2. That is, 2 squares, or 4, or 8, or 16, etc. All the 1s should be "covered" at least once, meaning that every cell containing a 1 is in at least one encircled group.

   To obtain the simplest Boolean expression from a map, you should keep the number of groups as few as possible with each group containing as many 1s as possible (remember: 2^0, 2^1, 2^2, 2^3, etc. number of 1s in a group). It is OK to encircle any 1 more than once if that can reduce the number of groups.

5) Each group of encircled cells represents a term in the simplified equation. The simplified Boolean equation in SOP form is read from the map. In the following example, we will demonstrate how the K-mapping rules are applied.

Simplification

Example 1   Simplify this Boolean equation using the K-map method.

$$X = \overline{(C + D)} + \overline{A}C\overline{D} + A\,\overline{B}\,\overline{C} + \overline{A}\,\overline{B}\,C\,D + A\,C\,\overline{D}$$

Solution: The first term is not a product term. We can apply DeMorgan's theorem to convert it to a product term. Then, we have:

$$X = \overline{C}\,\overline{D} + \overline{A}C\overline{D} + A\,\overline{B}\,\overline{C} + \overline{A}\,\overline{B}\,C\,D + A\,C\,\overline{D}$$

There are four input variables, so the K-map should have four rows and four columns. Assign A and B as row variables, and C and D as column variables. The K-map layout is shown below.

	$\overline{C}\,\overline{D}$	$\overline{C}\,D$	$C\,D$	$C\,\overline{D}$
$\overline{A}\,\overline{B}$				
$\overline{A}\,B$				
$A\,B$				
$A\,\overline{B}$				

Map each product term into this K-map. There should be four 1s for the first product term since both A and B variables are missing. Furthermore, these four 1s must be in the first column, otherwise the A and B variables would not be cancelled. Two 1s should be used to represent the second term since one variable is missing from it. Note that several 1s may be assigned to the same cell. When finished, the K-map is as shown here:

	$\overline{C}\,\overline{D}$	$\overline{C}\,D$	$C\,D$	$C\,\overline{D}$
$\overline{A}\,\overline{B}$	1		1	1
$\overline{A}\,B$	1			1
$A\,B$	1			1
$A\,\overline{B}$	1	1		1

Grouping the 1s results in the final K-map shown here:

	$\overline{C}\,\overline{D}$	$\overline{C}\,D$	$C\,D$	$C\,\overline{D}$
$\overline{A}\,\overline{B}$	1		1	1
$\overline{A}\,B$	1			1
$A\,B$	1			1
$A\,\overline{B}$	1	1		1

Then, the simplified Boolean equation is:

$$X = \overline{D} + A\,\overline{B}\,\overline{C} + \overline{A}\,\overline{B}\,C$$

Relationship Between a K-Map and Its Truth Table

A truth table describes how an output is related to the inputs for all combinations of the inputs. Each 1 in the output column of a truth table represents a product term. For example, for a 4-input truth table, if the input ABCD = 1111 causes an output of 1, then ABCD is a product term. A 0 output does not produce a product term. The number of 1s in the output column is the number of product terms in the SOP equation.

Similarly, in a K-map the number of squares is equal to the number of input combinations. Each 1 in a cell corresponds to a full product term in the Boolean equation. Since each cell in the K-map corresponds to a row in the truth table, a truth table can be divided into sections such that each section corresponds to a row (or a column) in its K-map. This is illustrated in the following example.

Truth Table Simplification

Example 2

Given the following truth table, simplify the Boolean equation represented by this truth table using the K-map method.

Inputs					Output		
A	B	C	D		X		Row #
0	0	0	0		0		1st
0	0	0	1		1		"
0	0	1	0		1		"
0	0	1	1		0		"
0	1	0	0		1		2nd
0	1	0	1		1		"
0	1	1	0		0		"
0	1	1	1		0		"
1	0	0	0		0		4th
1	0	0	1		1		"
1	0	1	0		1		"
1	0	1	1		0		"
1	1	0	0		1		3rd
1	1	0	1		0		"
1	1	1	0		0		"
1	1	1	1		0		"

Table 6.1 Truth Table for Example 1.2

The Boolean equation represented by the above truth table is:

$$X = \overline{A}\,\overline{B}\,\overline{C}D + \overline{A}\,\overline{B}C\overline{D} + \overline{A}B\,\overline{C}\,\overline{D} + \overline{A}B\overline{C}D + A\overline{B}\,\overline{C}D + A\overline{B}C\overline{D} + AB\overline{C}\,\overline{D}$$

87

The following K-map corresponds to the Boolean equation:

	$\overline{C}\,\overline{D}$	$\overline{C}\,D$	$C\,D$	$C\,\overline{D}$
$\overline{A}\,\overline{B}$		1		1
$\overline{A}\,B$	1	1		
$A\,B$	1			
$A\,\overline{B}$		1		1

Using the groupings shown on the K-map, the simplified Boolean equation is:

$$X = \overline{A}\,B\,\overline{C} + B\,\overline{C}\,\overline{D} + \overline{B}\,\overline{C}\,D + \overline{B}\,C\,\overline{D}$$

PROCEDURE:

1)   For the truth table shown below:

   a)   Write the Boolean equation represented by this truth table in the sum-of-products form without simplification.

   b)   Build a logic circuit for this Boolean equation in the **Schematic Editor** window using AND, OR, and NOT gates.

   c)   Implement the design and download the JED file onto your target board.

   d)   Verify that the operation of the circuit matches the truth table.

Inputs					Output		VERIFIED	
A	B	C	D		X		YES	NO
0	0	0	0		1			
0	0	0	1		1			
0	0	1	0		0			
0	0	1	1		0			
0	1	0	0		1			
0	1	0	1		0			
0	1	1	0		1			
0	1	1	1		0			
1	0	0	0		1			
1	0	0	1		1			
1	0	1	0		0			
1	0	1	1		0			
1	1	0	0		0			
1	1	0	1		1			
1	1	1	0		0			
1	1	1	1		0			

Table 6.2 Truth Table Showing Result of the Experiment

Checked by _____ Date _____

2)   Simplify the Boolean equation in step 1 (a) using the K-map method.

3)   Create a logic design for the simplified Boolean equation (sum-of-products form) in step 2 using AND, OR, and NOT gates in the **Schematic Editor** window.

4)   Implement the design obtained in step 3 and download the JEDEC file onto your target board.

5)   Verify that the output of the new circuit gives the same truth table as the original circuit of step 1.

Checked by _____ Date _____

QUESTIONS:

1)   Use the K-map method to simplify the following equation:

$$X = \overline{A}\,\overline{B}\,\overline{C}\,\overline{D} + \overline{A}\,B\,\overline{C}\,\overline{D} + \overline{A}\,B\,C\,D + A\,\overline{B}\,C\,\overline{D} + A\,B\,C\,D$$

2)   Write the simplified Boolean equation for the following Karnaugh map. Then, draw the logic circuit using AND, OR, and NOT gates for the simplified equation.

	$\overline{C}\,\overline{D}$	$\overline{C}\,D$	$C\,D$	$C\,\overline{D}$
$\overline{A}\,\overline{B}$	1			1
$\overline{A}\,B$	1	1		1
$A\,B$				
$A\,\overline{B}$	1			1

USE FOR NOTES OR AS A WORK-SHEET

NAME _____ **DATE** _____

CPLD EXPERIMENT 7:

Binary Adders

OBJECTIVES:

- Design a 1-bit full adder based on its truth table.
- Demonstrate modular design and hierarchy by constructing a 4-bit adder using four 1-bit full adder modules.
- Verify the function of the 4-bit adder on the target board.

MATERIALS:

- Xilinx Foundation Software, student or professional edition V1.5 or higher.
- IBM or compatible computer with Pentium processor or equivalent, 64 M-byte RAM or more, and 3 G-byte or larger hard drive.
- PLDT-1 board by RSR Electronics Inc., XS95 board and XStend Board by XESS Corp., or a similar board with an XC95108 device.

DISCUSSION:

Addition and subtraction are two essential arithmetic functions performed by computers and other digital systems. It is therefore important to understand how to design a circuit to perform such functions. However, since subtraction is done by adding the 2s complement of a number, we will only need to design one adder circuit to perform both operations. An adder can be 1 or more bits. A 4-bit adder can add two 4-bit unsigned binary numbers. If larger binary numbers are to be added, an adder with more bits is needed. Let's observe what happens when adding two 4-bit binary numbers with pencil and paper:

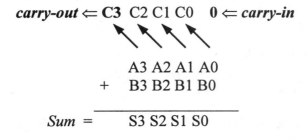

$$\textit{carry-out} \Leftarrow \textbf{C3} \; C2 \; C1 \; C0 \quad 0 \Leftarrow \textit{carry-in}$$

$$
\begin{array}{r}
A3 \; A2 \; A1 \; A0 \\
+ \quad B3 \; B2 \; B1 \; B0 \\
\hline
\textit{Sum} = \quad S3 \; S2 \; S1 \; S0
\end{array}
$$

The symbols [A3 A2 A1 A0] and [B3 B2 B1 B0] represent addend and minuend, respectively. C0 is the carry bit generated by adding bits A0 and B0. C1 is the carry bit generated from the addition of C0, A1, and B1. C2 and C3 are generated in the same manner, with C3 the carry-out. The column containing A0 and B0 (the least significant bits of addend and minuend) allows for a carry-in from a previous addition, for this example we set it to 0. Each column adds three bits. The implementation of the above process in hardware is called a *1-bit full adder*.

When we perform this addition, we will start from the least significant bit, and then we will push the process left one bit at a time. This means that a 1-bit full adder is the basic element of a 4-bit adder and four such elements are needed to construct a 4-bit adder.

The 1-Bit Full Adder

From the previous discussion we know that a 1-bit full adder should have three inputs: carry input (C_{in}), addend (A), and minuend (B). We can determine the number of output bits by looking at any column in the addition process, say, the column containing C0, A1 and B1. Assume all three bits are 1. Then the result is 3 which, in binary, is 11. The sum requires two bits but S1 can be only one bit, so there must be a carry to the next column. Each column will produce a sum bit and a carry output to the next more significant bit position. So the circuit for the 1-bit full adder should have two outputs: sum bit (S) and carry output (C_{out}) Below Table 7.1 shows the truth table for the 1-bit full adder:

Inputs				Outputs	
C_{in}	A	B		C_{out}	S
0	0	0		0	0
0	0	1		0	1
0	1	0		0	1
0	1	1		1	0
1	0	0		0	1
1	0	1		1	0
1	1	0		1	0
1	1	1		1	1

Table 7.1 Truth Table for a 1-Bit Full Adder

Using Boolean algebra, we can derive the following two equations for the sum bit and the carry output bit:

$$S = A \oplus B \oplus C$$

$$C_{out} = C_{in}(A + B) + AB$$

The above two equations can be implemented using a 3-input XOR gate, two 2-input AND gates, and two 2-input OR gates.

Section II. Building a 4-Bit Adder Using Full Adder (FA) Modules

In this part of the experiment, we will show how to perform the modular design by building the 4-bit adder using four 1-bit full adder modules.

1)   Create a new project called ***ADD4***. Make the selections and editor boxes in your **New Project** window as shown in Figure 7.3. Then, click on the **OK** button.

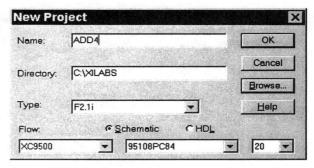

Figure 7.3 New Project window for project ***ADD4***.

2)   In the **Project Manager** window, choose **Document → Add** from the menu to add the C:\Xilabs\FA_1BIT\FA_1BIT.SCH file to the current project.

3)   Open the FA_1BIT.SCH file by double-clicking on the schematic name.

4)   Bypass all the IBUF and OBUF buffers. This means, connect the input and output of each buffer with a wire. Then, remove all the IBUF and OBUF buffers. Notice that you have to connect the bypassing wire before removing the buffer. Otherwise, the input and output terminals will disappear. The reason for removing the IBUFs and OBUFs is that some I/O pins, such as C0, C1, and C2, from the modules will not need to be assigned to CPLD pins. Figure 7.4 shows the logic to be encapsulated in the 1-bit full adder module.

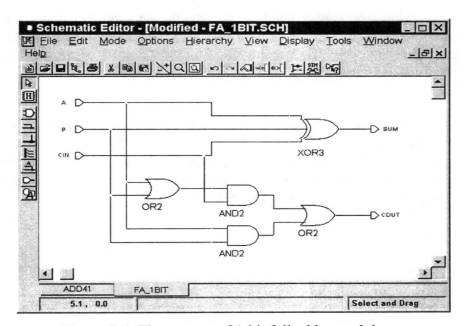

Figure 7.4 The content of 1-bit full adder module.

97

5)   To use this 1-bit full adder module in the 4-bit adder circuit, we need to create a symbol for it. Choose **Hierarchy → Create Macro Symbol from Current Sheet** from the menu. The **Create Symbol** window should appear.

6)   In the **Create Symbol** window, type *ADD1*, *1-bit full adder* in the **Symbol Name** and **Comment** editor boxes (You can name the symbol anything you want and write any meaningful comment.) as shown in Figure 7.5. Then, click on the **OK** button.

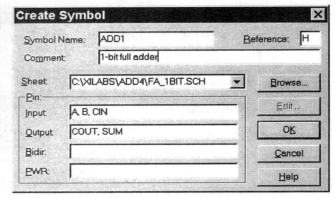

Figure 7.5   Create Symbol window for step 6.

7)   The message window shown in Figure 7.6 will pop up. The symbol *ADD1* has been saved into library. Since you have finished creating this symbol and you do not want to edit it at this time, click the **NO** button.

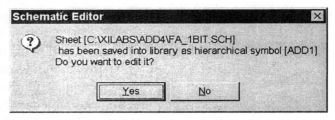

Figure 7.6   The message window for step 7.

8)   In your **Schematic Editor** window, you should see a new schematic sheet named ADD41.SCH. Open the **SC Symbol** window and locate ADD1 module (the 1st one on the list) Click on the module, drag and drop it on the schematic sheet. Make three copies of this symbol and line them up vertically.

9)   From the **SC Symbol** window, get nine IBUFs for the two 4-bit inputs (A3A2A1A0 and B3B2B1B0), and the carry-in (CIN) input, five OBUFs for the sum bits S3S2S1S0, and the carry-out (COUT) output. Add I/O terminals for all the buffers and name them as shown in Figure 7.7. Connect all necessary wires. Your schematic design should now look similar to Figure 7.7.

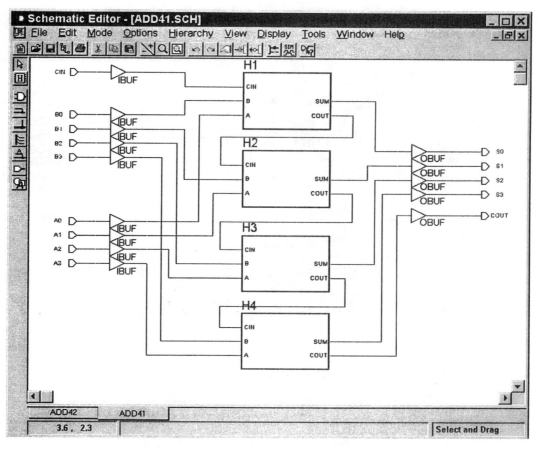

Figure 7.7 Schematic design of the 4-bit adder after step 9.

10) Save the schematic design. Go through **Options → Create Netlist, Options → Integrity Test** and **Options → Export Netlist...** from the menu. Then, close the **Schematic Editor** window.

11) If the PLDT-1 board is to be used to test this circuit, we need nine switches. There are only eight toggle switches on this board. To remedy this, we can connect one of the four columns on the TIE BLOCK B1, say, pin 12 (1st column) to GND during testing so that we get the CIN input 0. Following Tables 7.3 and 7.4, assign the CPLD pins to the I/O buffers. Note that LED1 through LED5 are used to display the SUM. Figure 7.8 shows the schematic design with pin assignment.

Input	A0	A1	A2	A3	B0	B1	B2	B3	CIN
Pin #	5	11	7	6	70	66	71	72	12

Table 7.3 Input Pin Assignment for the 4-Bit Adder

Output	S0	S1	S2	S3	S4
Pin #	39	40	41	43	44

Table 7.4 Output Pin Assignment for the 4-Bit Adder

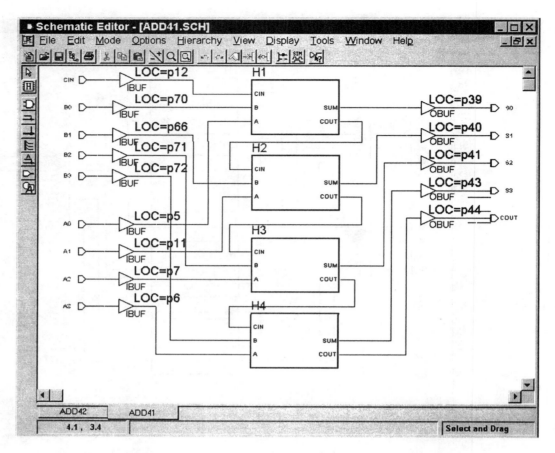

Figure 7.8 Schematic with pin assignment for the 4-bit adder.

12) Save the schematic design again. Repeat **Options → Create Netlist, Options → Integrity Test** and **Options → Export Netlist...** from the menu. Then, close the **Schematic Editor** window.

13) Implement this design and then program the XS9500 CPLD on your target board.

14) Remember to connect pin 12 (TIE BLOCK, column 1) to GND on the PLDT-1 board. Test the circuit design by giving different combinations of the inputs and observing the LEDs 1 through 5. LED1 is the most significant bit and LED5 is the least significant bit of the sum. Toggle switches 1 to 4 represent A3, A2, A1, and A0, respectively. Toggle switches 5 to 8 represent B3, B2, B1, and B0, respectively. Complete the following table by flipping the eight toggle switches on your target board. Note that the total combinations of the inputs are $2^8 = 64$. We only sample some of the input combinations in Table 7.5.

CPLD EXPERIMENT 8:

Decoders & Applications

OBJECTIVES:

- Examine the function of a decoder.
- Design and test a 2-to-4 decoder with active-low outputs using VHDL HDL .
- Design and simulate a BCD decoder design with active-high outputs in HDL.

MATERIALS:

- Xilinx Foundation Software, student or professional edition V1.5 or higher.
- IBM or compatible computer with Pentium processor or equivalent, 64 M-byte RAM or more, and 3 G-byte or larger hard drive.
- PLDT-1 board by RSR Electronics Inc., XS95 board and XStend Board by XESS Corp., or a similar board with an XC95108 device.

DISCUSSION:

So far, we have used Xilinx's **Schematic Editor** to create combinational circuits. But the Xilinx Foundation Series Tools also contain Hardware Description Languages (HDLs) as an alternative method for creating combinational circuits. Xilinx Foundation software supports three types of HDL: ABEL, VHDL, and Verilog. We will use VHDL in this experiment. VHDL is an acronym for VHSIC (Very High-Speed Integrated Circuit) HDL. VHDL is an industry-standard HDL, and can be used to model a digital system at many levels of abstraction ranging from the algorithmic level to the gate level. VHDL code files have a sub-field with the suffix vhd. In the VHDL design, the netlist is extracted from the .vhd file, whereas, in a schematic design the netlist is extracted from the .sch file.

In the Xilinx schematic design tool, symbols for a 2-to-4 decoder, a 3-to-8 decoder, and a BCD decoder, among others, can be obtained directly from the **SC Symbol** list. In this experiment we will build a decoder from basic logic operations. The purpose of this experiment is to demonstrate how to use an HDL tool, such as VHDL, to design combinational circuits. We will also see how to design decoders.

We will start with a simple 2-to-4 decoder. We will show the steps for entering VHDL code for the decoder and then synthesizing the netlist from the VHDL code. We also will show how to make pin selections using VHDL. The rest of the steps, such as implementation, simulation, and programming a target board, are exactly the same as we showed in previous experiments. We will then design a BCD decoder using VHDL and run it in simulation.

The 2-to-4 Decoder

Decoders are used often in computer and communication circuits. A decoder activates one of its outputs depending on the binary number present on its input. If the number of data inputs is N, the maximum number of outputs for a decoder is 2^N since there are 2^N possible combinations of the inputs. For a full decoder, there is an output for each combination of the inputs. For example, a 2-to-4 (or 1-out-of 4) decoder has 2 data inputs and hence $2^2 = 4$ outputs. A partial decoder does not use all possible input combinations, and so has fewer outputs. The truth table shown in Table 8.1 describes the functionality of a 2-to-4 decoder with active-low outputs.

Inputs			Outputs			
A	B		O_0	O_1	O_2	O_3
0	0		0	1	1	1
0	1		1	0	1	1
1	0		1	1	0	1
1	1		1	1	1	0

Table 8.1 The Truth Table for a 2-to-4 Decoder with Active-Low Outputs

If we examine the truth table closely, we find that there is only one 0 in each output column. So there is no need to use Boolean algebra or a K-map; we can get the equations by just seeing what combination of inputs produces the 0 in each column. The equations are:

$$\overline{O_0} = \overline{A}\,\overline{B}$$

$$\overline{O_1} = \overline{A}\,B$$

$$\overline{O_2} = A\,\overline{B}$$

$$\overline{O_3} = A\,B$$

The schematic diagram for this decoder is shown in Figure 8.1.

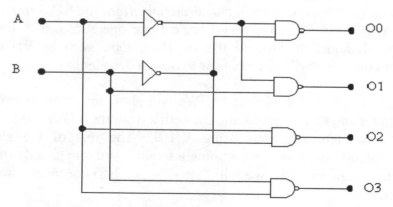

Figure 8.1   The 2-to-4 decoder with active-low outputs.

104

The BCD Decoder

Some decoders do not utilize all the 2^N possible input codes but only certain ones. An example is the decoder used in converting a binary-coded decimal (BCD) number to a decimal one.   The following truth table (Table 8.2) describes the function of the BCD decoder. The inputs, outputs, and enable (EN) are all active high.

	Inputs					Outputs									
EN	A	B	C	D	n	O_0	O_1	O_2	O_3	O_4	O_5	O_6	O_7	O_8	O_9
1	0	0	0	0	0	1	0	0	0	0	0	0	0	0	0
1	0	0	0	1	1	0	1	0	0	0	0	0	0	0	0
1	0	0	1	0	2	0	0	1	0	0	0	0	0	0	0
1	0	0	1	1	3	0	0	0	1	0	0	0	0	0	0
1	0	1	0	0	4	0	0	0	0	1	0	0	0	0	0
1	0	1	0	1	5	0	0	0	0	0	1	0	0	0	0
1	0	1	1	0	6	0	0	0	0	0	0	1	0	0	0
1	0	1	1	1	7	0	0	0	0	0	0	0	1	0	0
1	1	0	0	0	8	0	0	0	0	0	0	0	0	1	0
1	1	0	0	1	9	0	0	0	0	0	0	0	0	0	1
1	1	0	1	0	inv	0	0	0	0	0	0	0	0	0	0
1	1	0	1	1	inv	0	0	0	0	0	0	0	0	0	0
1	1	1	0	0	inv	0	0	0	0	0	0	0	0	0	0
1	1	1	0	1	inv	0	0	0	0	0	0	0	0	0	0
1	1	1	1	0	inv	0	0	0	0	0	0	0	0	0	0
1	1	1	1	1	inv	0	0	0	0	0	0	0	0	0	0
0	x	x	x	x		0	0	0	0	0	0	0	0	0	0

Table 8.2 Truth Table for the BCD Decoder with Active-High Outputs

As shown in the above truth table, when the enable is active (EN = 1) one of the ten outputs can be selected; but when EN = 0, no output will be activated regardless of the input ('x' here means "don't care"). For the invalid BCD inputs (1010 through 1111 and designated **inv**), no output will be selected. The Boolean equations can be derived the same way as we did previously for the 2-to-4 decoder. There are ten Boolean equations for the ten outputs:

$$O_0 = EN \bullet (\overline{A}\ \overline{B}\ \overline{C}\ \overline{D}), \quad O_1 = EN \bullet (\overline{A}\ \overline{B}\ \overline{C}\ D), \ \ldots, \quad O_9 = EN \bullet (A\ B\ C\ D)$$

The schematic diagram can be drawn easily once we have the Boolean equations. We can use four NOT gates and ten 5-input AND gates to implement the ten equations. The schematic diagram is left for you to draw in the questions section.

PROCEDURE:

Section I. Design a 2-to-4 Decoder Using VHDL

1) Create a directory for storing projects in HDL mode, and name it XI_HDL.

2) In your **Project Manager** window, choose **File → New Project** from the menu. Or, in the **Getting Started** window, click on the **Create a New Project** radio button and then click on the **OK** button.

3) In the **New Project** window, click on the **HDL** radio button. Once you select HDL mode, the chip family, chip part number, and device speed drop-down menus will disappear. Browse or type to get *C:\XI_HDL* in the **Directory** editor box, and type *DECO24* in the **Name** editor box. Then, click on the **OK** button.

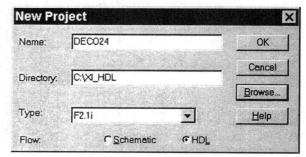

Figure 8.2 New Project window for the *DECO24* project.

4) Your **Project Manager** window should now look like Figure 8.3 below. Comparing this **Project Manager** window with the one in schematic mode, there is an additional button in the right pane labeled SYNTHESIS.

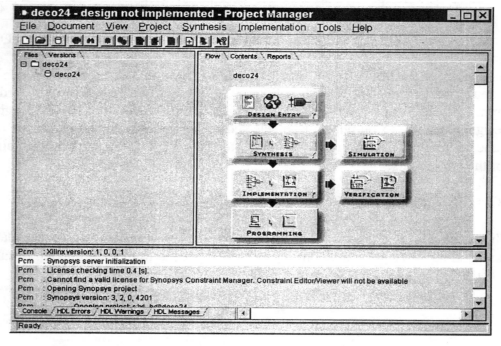

Figure 8.3 Project Manager window after step 4.

5) In the **Project Manager** window, choose **Tools → Design Entry → HDL Editor** or click on the 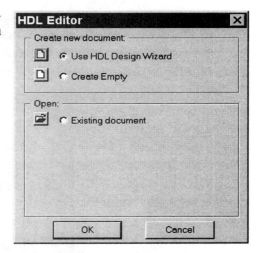 button on the **Design Entry** button.

6) The **HDL Editor** window pops up. Click on the **Use HDL Design Wizard** radio button. Then, click on the **OK** button.

Figure 8.4 HDL Editor window.

7) You should be in the **Design Wizard** window. In this window, click on the **Next** button to go to the **Design Wizard-Language** window.

8) In the **Design Wizard-Language** window, select the **VHDL** radio button and then click on the **Next** button to move to the **Design Wizard-Name** window.

9) In the **Design Wizard-Name** window, enter *DECO2_4.vhd* as the file name. Then, click on the **Next** button.

10) Now, you should be in the **Design Wizard-Ports** window. It should be blank at this point. We will add the inputs and outputs in this window for our design.

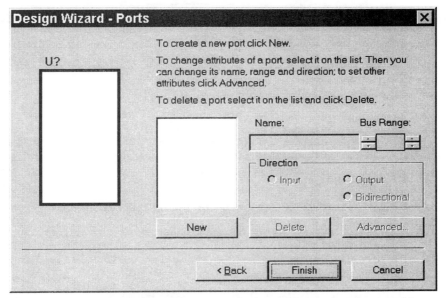

Figure 8.5 Design Wizard - Ports window before entries of I/O.

11) To add an input for the decoder, click on the **New** button and enter *A* (one of the inputs for the 2-to-4 decoder) in the **Name** field. Then, click on the **Input** radio button to specify the port to be an input. Repeat the same process for input B. To add an output, click on the **New** button and type *O0* in the **Name** field. Then, click on the **Output** radio button to set this port to be an output. Repeat the same process for the outputs *O1*, *O2*, and *O3*. Your **Design Wizard - Ports** window should now resemble Figure 8.6 below.

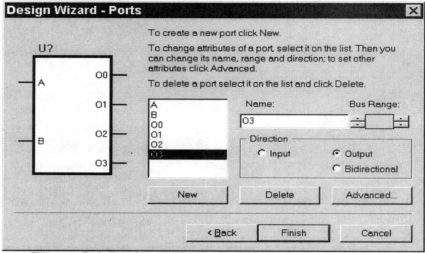

Figure 8.6 Design Wizard - Ports window after step 11.

12) Click on the **Finish** button in the **Design Wizard - Ports** window to go to the **HDL Editor** window. Notice that the **HDL Editor** window already has the template of the VHDL code for the *DECO2_4* logic as shown in Figure 8.7.

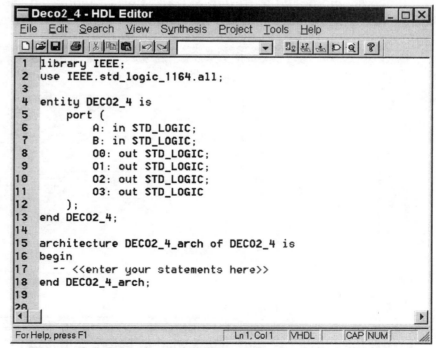

Figure 8.7 Template of VHDL codes for *Deco2_4* logic.

108

In the above .vhd file, these are keywords: *library, use, all, entity, is, port, in, out, end, architecture, of,* and *begin.* They are colored red on your screen. The keywords can be in either lower case or upper case.

The first line of the file is similar to the #include directive in the C programming language in that it   specifies the libraries used in the design. The IEEE library allows us to access the macros, definitions, and functions. A library can be further divided into packages. Each package contains a group of features in a certain area.

The second line specifies that this design will have access to *all* the features in the STD_LOGIC_1164 package of the IEEE library.

Lines 4 through 13 define the interface to the *DECO2_4* circuit. All the inputs and outputs are declared here. External circuitry can access the functions of the circuit through those inputs and outputs. Note that we have to follow the specific format of the syntax for the interface definition.

Line 15 is the start of the architecture part for the *DECO2_4* circuit. It is in this part that we will enter the statement code to describe our decoder circuit.

13)   As stated in the discussion section of this experiment, we need to add four statements in the **HDL Editor** window to describe the functionality of the 2-to-4 decoder circuitry.

Enter the following four statements starting at line 18:

O0 <= *not* (*not* A *and not* B);
O1 <= *not* (*not* A *and* B);
O2 <= *not* (A *and not* B);
O3 <= *not* (A *not* B);

The keywords *not, and*, and *or* are used to define the logic operations of the circuit. Note: VHDL does not allow us to write complemented outputs (such as *not* O1). So, the right hand side of these Boolean equations are complemented to obtain uncomplemented outputs. Figure 8.8 shows the VHDL code after the addition of the above four statements.

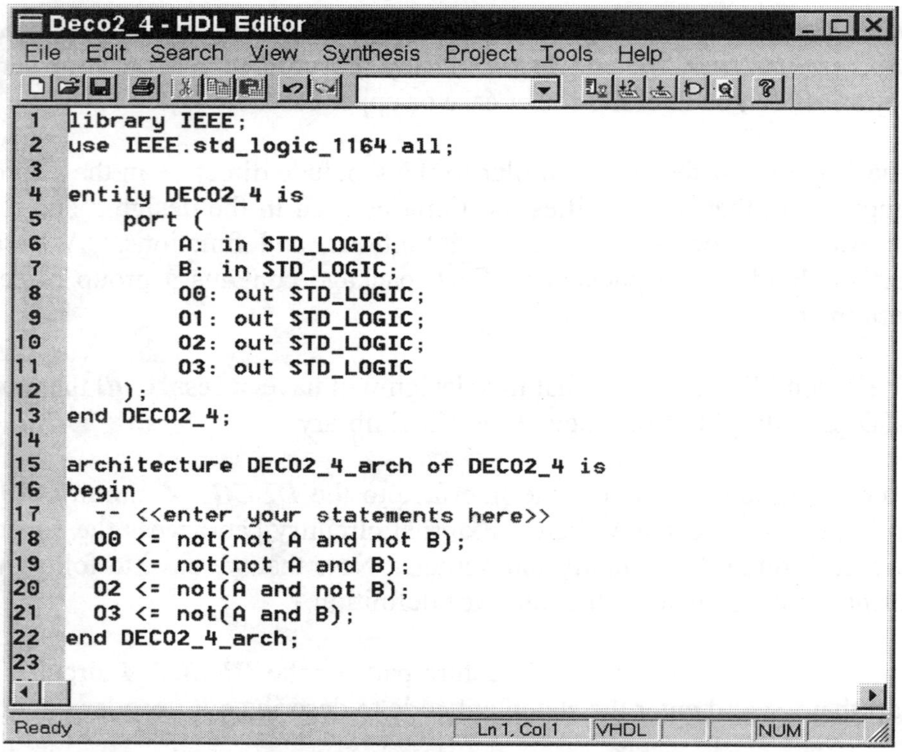

```
▆Deco2_4 - HDL Editor                                    _ □ ✕
File   Edit   Search   View   Synthesis   Project   Tools   Help
▢☞▣│🖨│✂▦▦│↶↷│            ▾│ ▨▨▨▨▢◧?│

 1  library IEEE;
 2  use IEEE.std_logic_1164.all;
 3
 4  entity DECO2_4 is
 5      port (
 6          A:  in STD_LOGIC;
 7          B:  in STD_LOGIC;
 8          O0: out STD_LOGIC;
 9          O1: out STD_LOGIC;
10          O2: out STD_LOGIC;
11          O3: out STD_LOGIC
12      );
13  end DECO2_4;
14
15  architecture DECO2_4_arch of DECO2_4 is
16  begin
17    -- <<enter your statements here>>
18    O0 <= not(not A and not B);
19    O1 <= not(not A and B);
20    O2 <= not(A and not B);
21    O3 <= not(A and B);
22  end DECO2_4_arch;
23
◀│                                                          ▶│
Ready                             Ln 1, Col 1   VHDL        NUM
```

Figure 8.8 HDL Editor window after adding the four statements.

14) Next, let's check if we made any syntax mistakes in the statements entered. Choose **Synthesis → Check Syntax** from the menu. If there are no errors, a small window pops up saying **Check Successful**. If there are one or more syntax errors, the pop-up window will say **Errors Found** and the lines with errors will be highlighted. A small red arrow will point at the line where the first error occurs. Also, you will see an error message on the bottom of this window.

There are error-free VHDL code examples available in the Xilinx Series Tools. To access them, select **Tools → Language Assistant** from the menu. The **Language Assistant – VHDL** window will appear. You can study any particular topic listed in this window by clicking on the + symbols and the specific topic. You can also cut and paste the example into your VHDL code file, and then edit as needed.

15) Save this design by selecting **File → Save** in the **HDL Editor** window.

16) To add this file to the project, select **Project → Add to Project** from the menu. Save this file again and exit to the **Project Manager** window.

17) Recall that we did not write any code in the VHDL file to make an I/O pin assignment. In fact, this information cannot be included in a .vhd file. We have to put it in the .ucf file (user-constraint file). Right-click on the *deco24* project name in the left pane of the **Project Manager** window. In the sub-menu that appears, choose **Edit Constraints** to open the **Report Browser** window.

110

18) In the **Report Browser** window, add the following 6 statements:

```
NET  A      LOC=P6;
NET  B      LOC=P7;
NET  O0     LOC=P44;
NET  O1     LOC=P43;
NET  O2     LOC=P41;
NET  O3     LOC=P40;
```

As you can see, the inputs A and B are assigned to pins 6 and 7 (toggle switches 1 and 2). The outputs O0, O1, O2, and O3 use pins 44, 43, 41, and 40 (LEDs 1 through 4), respectively. Figure 8.9 depicts the **Report Browser** for the deco24.ucf file. Note that in .ucf files, comment lines start with ' # '.

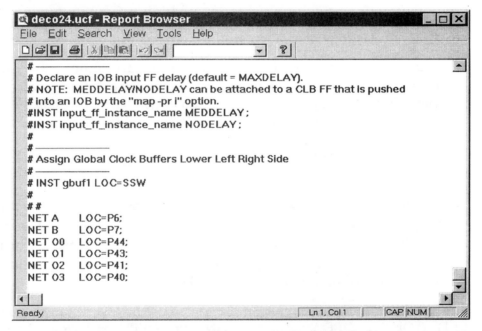

Figure 8.9 I/O pins for the 2-to-4 decoder are specified in *deco24.ucf*.

19) Save this file by choosing **File → Save**. Then choose **File → Exit** to go back to the **Project Manager** window.

20) In the **Project Manager** window, choose **Synthesis** → **Synthesize...** from the menu or click on the **Synthesis** button. Make sure that all the selections are the same as shown in the **Synthesis/Implementation settings** window in the right. Then, click on the **Run** button to complete the synthesis. When this process is finished, you will return to the **Project Manager** window. Both the **Design Entry** and **Synthesis** buttons are marked with green checks.

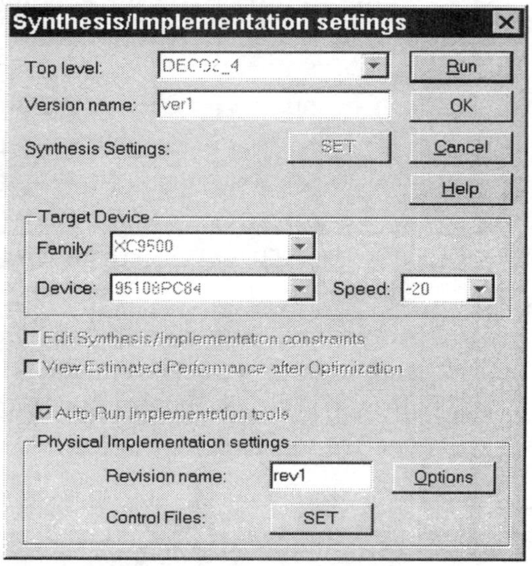

Figure 8.10
Synthesis/Implementation settings for the 2-to-4 decoder.

21) To implement the design, click on the **Implementation** button or choose **Implementation** → **Implement Design** from the menu. The **Synthesis/Implementation settings** window will appear. Click on the **Run** button to start the implementation. You will see the Flow Engine and the steps of the process. Click on the OK button in the message sub-window to go back to the **Project Manager** window. The Implementation button should now show a green checkmark (√).

22) We can now download the bitstream. Connect power to your target board and connect it to your PC. Click on the **Programming** button or choose **Tools** → **Device Programming** → **JTAG Programmer** from the menu. In the JTAG Programmer window, choose **Operations** → **Program** from the menu. It takes 5 to 30 seconds to complete the download process.

23) Test the 2-to-4 decoder circuit. In the truth-table below, record the status of the LEDs ("ON" or "OFF"). Note that "ON" and "OFF" denote logic 1 and 0 for the outputs. Is this table the same as we discussed at the beginning of this lab?

Inputs			Outputs			
A	**B**		**O_0**	**O_1**	**O_2**	**O_3**
SW 1	**SW 2**		**LED 1**	**LED 2**	**LED 3**	**LED 4**
L	L					
L	H					
H	L					
H	H					

Table 8.3 Test Results for the 2-to-4 Decoder

Checked by _____ Date _____

Section II. Design a BCD Decoder Using VHDL

Refer back to the discussion of the truth table and Boolean equations for the BCD decoder.

1)   Follow the steps 2 through 5 of section I above to create a new project in HDL mode. Name the project **DECOBCD**.

2)   Follow the steps 6 through 12 of section I. Type **DECO_BCD.vhd** in the **Design Wizard-Name** window. Enter the inputs (EN, A, B, C, D) and outputs (O0 to O9) in the **Design Wizard – Port** window. When you finish, your VHDL code template will be generated and shown in the **HDL Editor** window.

3)   In the **HDL Editor** window, enter the ten statements representing the ten Boolean equations for the BCD decoder following <<Enter your statements here>> as illustrated in Figure 8.11 below. To save time, type the first equation, then copy, paste, and edit to make the rest.

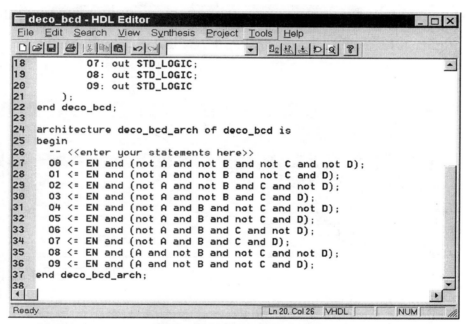

Figure 8.11   VHDL codes for the BCD decoder.

4)   Follow steps 14 to 16 from section I, check the syntax of the VHDL file, add the file to the project, and save the file.

5)   Refer to step 20 in section I to synthesize this VHDL file. Since we plan to simulate the BCD decoder, we do not need to assign CPLD pins to perform the implementation and programming.

6)   Refer to previous experiments for the steps to perform a simulation of the BCD decoder. The results should look similar to Figures 8.12 and 8.13. Note that data inputs A, B, C, and D are assigned to counter stimulators B3, B2, B1 and B0, respectively. The control input EN is tied to toggle switch "c".

113

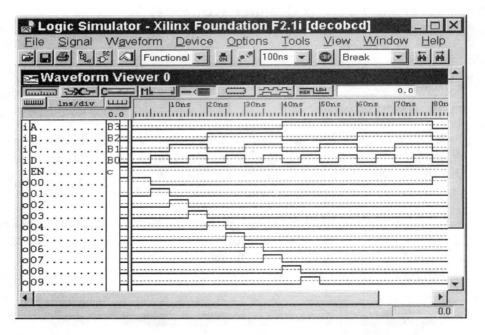

Figure 8.12 Timing waveforms for the BCD decoder when EN = 1.

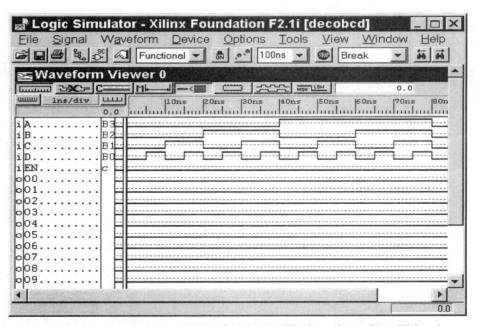

Figure 8.13 Timing waveforms for the BCD decoder when EN = 0.

7)   Compare your simulation results with Table 8.2 in the discussion section. Comment on how well the simulation verifies your expectations. Make notes here for future reference.

Check by _____   Date _____

QUESTIONS:

1)   For a 3-to-8 decoder with active-high outputs and an active-high enable line (EN):

a) List the truth table:

b) Write the Boolean equations:

c) Sketch the input and output timing waveforms for all input combinations.

2)   Name two applications for decoders.

3)   What use could an enable line have in a decoder?

4)   What is the TTL chip number for a 3-to-8 decoder with active-low outputs and enable lines?

5)   What is the TTL chip number for a BCD decoder with active-low outputs?

NAME _____ **DATE** _____

CPLD EXPERIMENT 9:
Encoders & Application to a 7-Segment Display Driver

OBJECTIVES:

- Examine the functions of encoders and their applications.
- Examine operating characteristics of a decimal-to-BCD priority encoder.
- Design an 8-to-3 line priority encoder in schematic mode and test the circuit on the target board.
- Create a macro for the 7-segment decoder with active-HIGH outputs using the VHDL HDL tool.
- Design a circuit in schematic mode utilizing a created macro to display a decimal digit on a 7-segment display.

MATERIALS:

- Xilinx Foundation Software, student or professional edition V1.5 or higher.
- IBM or compatible computer with Pentium processor or equivalent, 64 M-byte RAM or more, and 3 G-byte or larger hard drive.
- PLDT-1 board by RSR Electronics Inc., XS95 board and XStend Board by XESS Corp., or a similar board with an XC95108 device.

DISCUSSION:

As we saw in the previous experiment, a decoder identifies or detects a particular binary number or code. Encoding is the opposite process of decoding. An encoder has a number of input lines (up to 2^N), only one of which is activated at any given time, and produces the N-bit output code for the input selected. For example, consider an 8-to-3 encoder. When one of its eight inputs is activated, the output will be a 3-bit binary number (code) corresponding to that input.

If two or more inputs are activated at the same time, which one of the inputs should be encoded and reflected on the outputs? This is when priority is used in the encoder design. When multiple inputs are activated, priority specifies which input will get selected to produce the output code.

A common application of encoders is in the keyboards of calculators and computer systems to convert key-presses to binary numbers or to codes such as BCD or ASCII.

The 8-to-3 Line Encoder with Active-LOW Inputs

The 8-to-3 (octal-to-binary) encoder accepts eight input lines and produces a 3-bit output code corresponding to the activated input. Table 9.1 below describes the function of this encoder. Note the active-low inputs, as could be obtained from a keypad with normally-open contacts to ground.

Inputs									Outputs		
$\overline{A_0}$	$\overline{A_1}$	$\overline{A_2}$	$\overline{A_3}$	$\overline{A_4}$	$\overline{A_5}$	$\overline{A_6}$	$\overline{A_7}$		O_0	O_1	O_2
X	1	1	1	1	1	1	1		0	0	0
X	0	1	1	1	1	1	1		0	0	1
X	1	0	1	1	1	1	1		0	1	0
X	1	1	0	1	1	1	1		0	1	1
X	1	1	1	0	1	1	1		1	0	0
X	1	1	1	1	0	1	1		1	0	1
X	1	1	1	1	1	0	1		1	1	0
X	1	1	1	1	1	1	0		1	1	1

Table 9.1   Truth Table for the Octal-to-Binary Encoder

Looking at the truth table, we see that a LOW on any input will produce the binary code corresponding to that input. For instance, a LOW on A_5 produces the output 101, which is binary code for 5. For this example, the A_0 column is marked with "X" (don't care) because the outputs will be 000 when A_1 through A_9 are all high (not activated).

The Decimal-to-BCD Priority Encoder

Figure 9.2 is the truth table for a decimal-to-BCD priority encoder (such as the 74147 TTL chip). It has nine active-low inputs representing decimal numbers 1 through 9. The encoder produces the inverted BCD code corresponding to which of the nine inputs is activated.

Inputs									Outputs			
$\overline{A_1}$	$\overline{A_2}$	$\overline{A_3}$	$\overline{A_4}$	$\overline{A_5}$	$\overline{A_6}$	$\overline{A_7}$	$\overline{A_8}$	$\overline{A_9}$	$\overline{O_0}$	$\overline{O_1}$	$\overline{O_2}$	$\overline{O_3}$
1	1	1	1	1	1	1	1	1	1	1	1	1
X	X	X	X	X	X	X	X	0	0	1	1	0
X	X	X	X	X	X	X	0	1	0	1	1	1
X	X	X	X	X	X	0	1	1	1	0	0	0
X	X	X	X	X	0	1	1	1	1	0	0	1
X	X	X	X	0	1	1	1	1	1	0	1	0
X	X	X	0	1	1	1	1	1	1	0	1	1
X	X	0	1	1	1	1	1	1	1	1	0	0
X	0	1	1	1	1	1	1	1	1	1	0	1
0	1	1	1	1	1	1	1	1	1	1	1	0

Table 9.2   Truth Table for a Decimal-to-BCD Priority Encoder (74147)

Note the "don't-cares" (Xs) in the truth table. They imply that, if two inputs are activated simultaneously, only the highest data line is encoded. For example, if lines A_1 and A_5 are activated at the same time, A_5 will be encoded producing the output 1010 (which is 0101 inverted, or BCD 5). That's why it's called a "priority" encoder. Moreover, the implied decimal zero condition requires no inputs since zero is encoded when all nine data lines are at HIGH. For a more detailed discussion of encoder circuits, refer to your digital textbook.

The 7-Segment Display

A 7-segment display is composed of seven bars (commonly made with LEDs) that can be individually activated to emit light. Such a display can show digits from 0 to 9, as well as a few letters (A, b, C, d, E, F, H, L, P, S, U, Y), a minus sign (-) and a decimal point. For a common-cathode 7-segment LED display, the "common" input is connected to GND, and a HIGH on any segment-input will light up that segment. For a common-anode display, the common-input is connected to a HIGH, and a LOW on a segment-input lights the segment. Figure 9.1 (a) and (b) show how the seven segments are arranged.

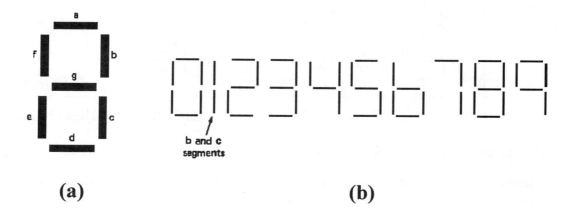

(a)   **(b)**

Figure 9.1 (a) 7-segment arrangement; **(b)** active segments for 0 to 9.

The 7-Segment Decoder

A 7-segment decoder is not available in the symbol library of the Xilinx software, so we will design one. We will design the decoder with active-high outputs for a common-cathode display. The lit display segments for each digit from 0 to 9 are given in the following truth table:

Inputs					Outputs						
D3	D2	D1	D0	N	a	b	c	d	e	f	g
0	0	0	0	0	1	1	1	1	1	1	0
0	0	0	1	1	0	1	1	0	0	0	0
0	0	1	0	2	1	1	0	1	1	0	1
0	0	1	1	3	1	1	1	1	0	0	1
0	1	0	0	4	0	1	1	0	0	1	1
0	1	0	1	5	1	0	1	1	0	1	1
0	1	1	0	6	1	0	1	1	1	1	1
0	1	1	1	7	1	1	1	0	0	0	0
1	0	0	0	8	1	1	1	1	1	1	1
1	0	0	1	9	1	1	1	0	0	1	1
1	0	1	0	inv	0	0	0	0	0	0	0
1	0	1	1	inv	0	0	0	0	0	0	0
1	1	0	0	inv	0	0	0	0	0	0	0
1	1	0	1	inv	0	0	0	0	0	0	0
1	1	1	0	inv	0	0	0	0	0	0	0
1	1	1	1	inv	0	0	0	0	0	0	0

Table 9.3 Truth Table for a 7-Segment Decoder with Active-High Outputs

Note that the outputs of this decoder are all LOW for invalid (inv) BCD codes.

Based on the above truth table, we can derive the Boolean equations for the outputs using K-maps or Boolean algebra:

$$a = \overline{D_3}D_1 + \overline{D_2}\,\overline{D_1}\,\overline{D_0} + D_3\overline{D_2}\,\overline{D_1} + \overline{D_3}D_2D_0$$

$$b = \overline{D_3}\,\overline{D_2} + \overline{D_2}\,\overline{D_1} + \overline{D_3}\,\overline{D_1}\,\overline{D_0} + \overline{D_3}D_1D_0$$

$$c = \overline{D_3}D_2 + \overline{D_2}\,\overline{D_1} + \overline{D_3}D_0$$

$$d = \overline{D_2}\,\overline{D_1}\,\overline{D_0} + \overline{D_3}\overline{D_2}D_1 + \overline{D_3}D_1\overline{D_0} + \overline{D_3}D_2\overline{D_1}D_0$$

$$e = \overline{D_2}\,\overline{D_1}\,\overline{D_0} + \overline{D_3}D_1\overline{D_0}$$

$$f = \overline{D_2}\,\overline{D_1}\,\overline{D_0} + \overline{D_3}D_2\overline{D_1} + \overline{D_3}D_2\overline{D_0} + D_3\overline{D_2}\,\overline{D_1}$$

$$g = \overline{D_3}\,\overline{D_2}D_1 + \overline{D_3}D_2\overline{D_1} + D_3\overline{D_2}\,\overline{D_1} + \overline{D_3}D_1\overline{D_0}$$

PROCEDURE:

Section I.   An 8-to-3 Line Encoder (74148)

1)   Create a new project called *enco83* in schematic mode. Your **New Project** window should look like Figure 9.2. Click on the **OK** button to go to the **Project Manager** window.

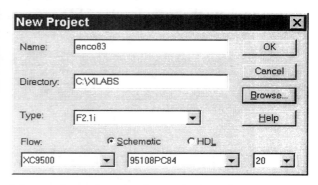

Figure 9.2 New Project window for *enco83*.

2)   In the **Project Manager** window, click on the **Schematic Editor** button on the **Design Entry** button to get into the **Schematic Editor** window. Create a schematic diagram similar to the one in Figure 9.3 below. Note that we used X74_148 from the **SC Symbol** window. I0 through I7 are the data inputs. EI (enable input) is a control input. A0 to A2 are the encoded data outputs. To cascade encoders for a multi-digit display, the EO of each stage is connected to the EI of the next stage (EI of first stage is held low). GS goes low whenever one of the data inputs goes low.

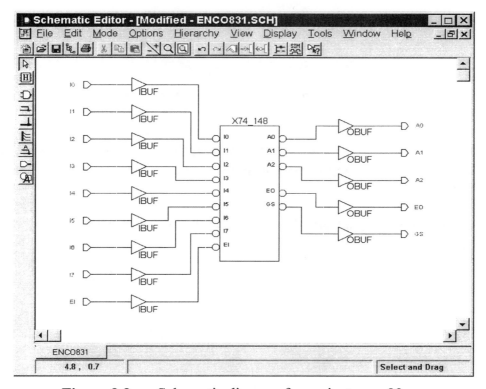

Figure 9.3   Schematic diagram for project *enco83*.

123

3)   Make the pin assignment for your design. For the PLDT-1 board, toggle switches 1 through 8 represent inputs I0, I1, ..., I7, respectively. The input EI is connected to pin 12, which is on column 1 of TIE BLOCK B1. The encoded data A2, A1 and A0 are displayed on LEDs 1, 2, and 3, respectively. LEDs 7 and 8 are used to show the logic levels of EO and GS, respectively.

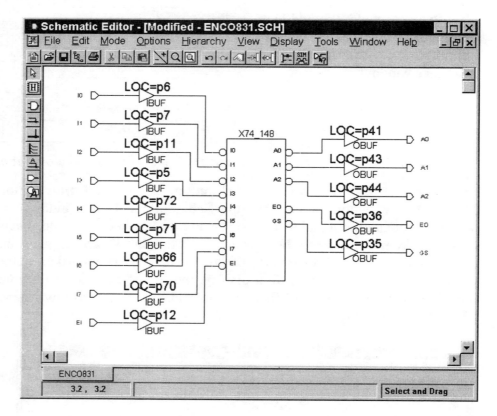

Figure 9.4   Pin assignments for the 8-to-3 encoder on the PLDT-1 board.

4)   Save the schematic design. Go through **Options → Create Netlist, Options → Integrity Test** and **Options → Export Netlist...** from the menu. Then, close the **Schematic Editor** window.

5)   Implement this design and then download the bitstream onto your target board.

6) Test the 8-to-3 line encoder and complete the function table below. Note that both the inputs and the outputs are active-low for this encoder. So an input is activated by a LOW logic level and the outputs are inverted binary numbers.

Inputs										Outputs				
$\overline{EI}$	$\overline{I_0}$	$\overline{I_1}$	$\overline{I_2}$	$\overline{I_3}$	$\overline{I_4}$	$\overline{I_5}$	$\overline{I_6}$	$\overline{I_7}$		$\overline{A_2}$	$\overline{A_1}$	$\overline{A_0}$	EO	$\overline{GS}$
H	X	X	X	X	X	X	X	X						
L	H	H	H	H	H	H	H	H						
L	X	X	X	X	X	X	X	L						
L	X	X	X	X	X	X	L	H						
L	X	X	X	X	X	L	H	H						
L	X	X	X	X	L	H	H	H						
L	X	X	X	L	H	H	H	H						
L	X	X	L	H	H	H	H	H						
L	X	L	H	H	H	H	H	H						
L	L	H	H	H	H	H	H	H						

Table 9.4 The 8-to-3 Line Encoder with Active-Low Inputs and Outputs

7) Based on what you observe from the completed table 9.4, how does EI function? Explain how the active-low outputs form an octal number. Comment on why the EO and GS outputs behave as they do.

Checked by _____ Date _____

125

Section II.   Creating a Macro for a 7-Segment Decoder

In this section of the experiment, we will build a macro for the 7-segment decoder with active-high outputs in HDL mode. We will save the design as a macro, and use it in Section III and in later experiments.

1)   Create a new project called **deco7seg** in HDL mode. Your **New Project** window should look like Figure 9.5. Click on the **OK** button to go to the **Project Manager** window.

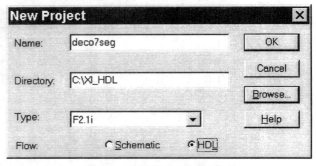

Figure 9.5 New Project window for **deco7seg**.

2)   If you do not remember how to start the design in HDL mode, refer back to CPLD experiment 8. In the **Design Wizard – Ports** window, make four inputs: D3, D2, D1, and D0 and seven outputs: a, b, c, d, e, f, and g. Click on the **Finish** button to go to the **HDL Editor** window.

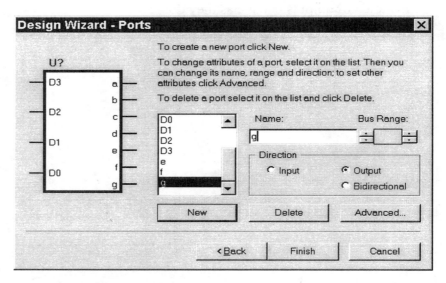

Figure 9.6 Design Wizard – Ports window for the **deco7seg** project.

126

3) In the **HDL Editor** window, add the seven statements representing the seven Boolean equations following <<enter your statements here>>. Figure 9.7 below shows the VHDL codes for our 7-segment decoder. Note that the .vhd file is saved under a different name (*deco7g.vhd*) from the project name. Check the syntax by selecting **Synthesis → Check Syntax** from the menu.

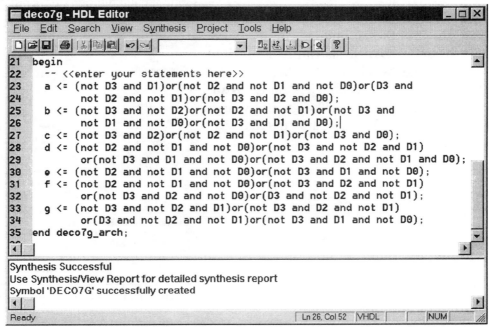

Figure 9.7 VHDL codes for the 7-segment decoder with active-high outputs.

4) In the **HDL Editor** window, choose **Project → Create Macro** from the menu. Make the selections shown in Figure 9.8 in the **Set initial target** window. Then, click on the **OK** button.

Figure 9.8 Set initial target window for the *deco7g* macro.

5) If there are no syntax errors and no obvious logic conflicts (such as two different Boolean equations represent the same output), you will see the pop-up window indicating the symbol was created successfully. Click on the **OK** button.

Figure 9.9 Pop-up window shows the macro was created successfully.

127

6) The macro *deco7g* is now created. Upon returning to your **Project Manager** window, get back to the **HDL Editor** window and choose **Project → Add to Project** from the menu. This is to include this macro in the project *deco7seg* so that we can perform a simulation to check the correctness of the design. Exit to the **Project Manager** window.

7) Synthesize the macro by clicking on the **Synthesis** button. Make sure your **Synthesis/Implementation settings** window looks the same as Figure 9.10. Then, click on the **Run** button. Both the Design Entry and Synthesis buttons are marked by green checks.

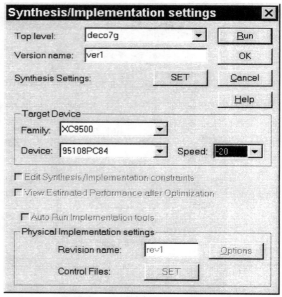

Figure 9.10 Settings for synthesizing the macro *deco7g*.

8) Run a simulation to check if the design functions properly. Your results should be like Figure 9.11 below. You may use this macro in any circuit from now on.

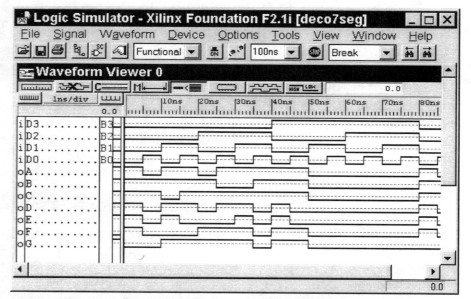

Figure 9.11 Simulation results for the 7-segment decoder.

Checked by_____ Date _____

128

Section III. Decimal Numbers on a 7-Segment Display

In this section of the experiment, we will build a circuit that can take a decimal input and encodes it to a BCD, which is then decoded to drive a 7-segment display.

1)   Create a new project named *dec2bcd* in schematic mode and make your **New Project** window look the same as Figure 9.12. Then, click on the **OK** button.

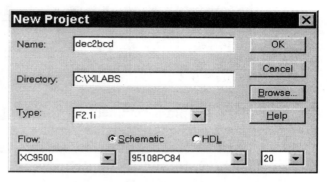

Figure 9.12 New Project window for *dec2bcd*.

2)   The macro *deco7g* is used in the *decc7seg* project. To bring it into the **SC Symbol** list for our new project, select **Tools → Project Libraries** from the menu in the **Schematic Editor** window. In the **Project Libraries** window, browse through the project list on the left and highlight *deco7seg*, and click on the **Add>>** button. Then, click on the **Close** button to go back to the **Schematic Editor** window.

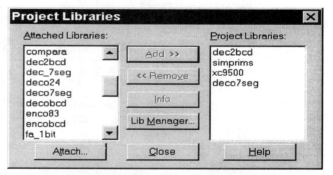

Figure 9.13 Project Libraries window.

3)   The *deco7g* macro is now included in the **SC Symbol** window and is ready for us to retrieve. Draw the schematic diagram shown in Figure 9.14. Note that inverters are used since the X74_147 decimal-to-BCD encoder has active-low outputs and our *deco7g* macro has active-high inputs.

4) There are nine inputs. On the PLDT-1 board, eight inputs can be from the eight toggle switches and one input can be connected to a column on TIE BLOCK B1. The outputs go to the 7-segment display. The PLDT-1 uses specific CPLD pins for the 7-segment display as listed in Table 9.5 below (dp is the decimal point).

LED Segment	a	b	c	d	e	f	g	dp
CPLD Pin #	15	18	23	21	19	14	17	24

Table 9.5 CPLD Pins Assigned to the 7-Segment Display on the PLDT-1

5) Follow the pin assignment shown in Figure 9.14. Assign inputs I1 through I8 to toggle switches 1 through 8, respectively. Use a jumper wire to drive column 1 of TIE BLOCK B1 high or low for I9. Assign outputs A through G to pins 15, 18, 23, 21, 19, 14, and 17, respectively.

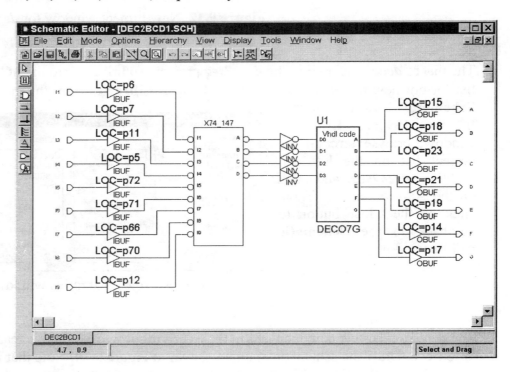

Figure 9.15 I/O pin assignment for the project **dec2bcd**.

6) Save this schematic design. Go through **Options → Create Netlist**, **Options → Integrity Test** and **Options → Export Netlist...** from the menu. Then, close the **Schematic Editor** window.

7) Implement the design and then program the 95108 chip on your target board.

8) Test your design on your target board. Note that you must use a jumper wire to connect pin 12 to GND while you are testing other decimal numbers since a floating input is undetermined in the CPLD chip. The encoder we are using is a priority encoder. What does the 7-segment show when no input is activated?

9) Describe the function of this circuit in your own words.

Checked by _____ Date _____

QUESTIONS:

1) Priority encoders are much more common than non-priority encoders. What do you think the reasons are for that?

2) Why is there no I_0 input in the decimal-to-BCD encoder?

3) If a common-anode 7-segment is used, what modifications must you make on the project design in Section III?   If you have more than one, list them all.

4) Name two applications for encoders.

CPLD EXPERIMENT 10:
Multiplexers and Demultiplexers

OBJECTIVES:

- Examine the functions of multiplexers (MUX) and demultiplexers (DEMUX).
- Create an 8-to-1 MUX in schematic mode and simulate the design.
- Create a 1-of-8 DEMUX in schematic mode and test the design on a target board.
- Compare the characteristics of the 1-of-8 DEMUX with the 3-to-8 decoder.

MATERIALS:

- Xilinx Foundation Software, student or professional edition V1.5 or higher.
- IBM or compatible computer with Pentium processor or equivalent, 64 M-byte RAM or more, and 3 G-byte or larger hard drive.
- PLDT-1 board by RSR Electronics Inc., XS95 board and XStend Board by XESS Corp., or a similar board with an XC95108 device.

DISCUSSION:

A multiplexer (MUX) is a logic circuit that channels two or more input data lines to one output data line. Hence, it is also called a data selector. The routing of a particular data input to the output is controlled by the SELECT (or ADDRESS) inputs. Generally speaking, a MUX has N select inputs (address bits), 2^N data inputs, and one data output. For example, an 8-to-1 MUX has eight data inputs, three select inputs, and one output.

Multiplexers are widely used in digital and data communication systems. They can perform data selection, data routing, operation sequencing, parallel-to-serial conversion, waveform generation, and logic-function generation. Multiplexers make it possible for several streams of digital data to be sent over one physical cable in a system called TDM (time-division-multiplexing) or TDMA (time-division-multiple-access).

A demultiplexer (DEMUX) performs the reverse operation of a MUX. A DEMUX is also called a data distributor. It selects a single data stream (channel) out of the several coming in and routes it to the appropriate output. The channel is chosen by putting its binary address on the select inputs. A DEMUX has one data input, N select inputs, and 2^N output lines. For instance, a 1-of-16 DEMUX has one data input, four select inputs, and 16 outputs.

The 8-to-1 MUX (74151)

The 8-to-1 MUX accepts one of the eight data inputs and passes the data to the output depending on the status of the SELECT lines. Table 10.1 describes the function of this MUX.

	Control Inputs				Output
$\overline{G}$	A	B	C		Y
1	x	X	X		0
0	0	0	0		D_0
0	0	0	1		D_1
0	0	1	0		D_2
0	0	1	1		D_3
0	1	0	0		D_4
0	1	0	1		D_5
0	1	1	0		D_6
0	1	1	1		D_7

Table 10.1 Truth Table for the 8-to-1 MUX (74151)

In the truth table, $\overline{G}$ is the gate enable input and A, B, and C are SELECT inputs, D_0 through D_7 are data inputs. Since the enable input is active-low, an input channel can be selected only when the enable line is LOW. An input channel is selected by placing its binary address on the inputs A, B, and C. For example, when ABC = 110, the output on Y will be the D_6 bitstream.

The 1-of-8 DEMUX

In CPLD Experiment 8, we examined decoders. Given a 3-to-8 decoder with an enable input, we had three data inputs, one enable input, and eight outputs. However, the same circuit can be used for a different function. Specifically, we can use the enable input as a data input, and the three data inputs as select inputs. The result is that the 3-to-8 decoder becomes a 1-of-8 demultiplexer, as illustrated in Figure 10.1 below.

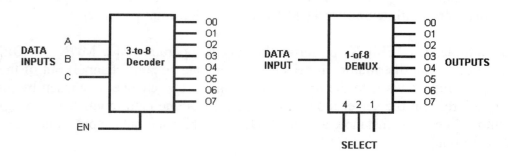

Figure 10.1 A decoder used as a DEMUX.

PROCEDURE:

Section I.   The 8-to-1 MUX

1)   Create a new project called **mux8_1** in schematic mode. Your **New Project** window should look like Figure 10.2. Click on the **OK** button to go to the **Project Manager** window.

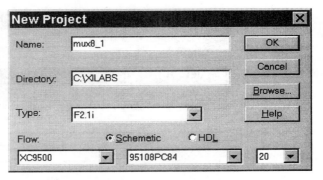

Figure 10.2 New Project window for **mux8_1**.

2)   In the **Project Manager** window, click on the **Schematic Editor** button in the **Design Entry** tab to get into the **Schematic Editor** window. Create a schematic diagram similar to Figure 10.3 below. Note that we used the **X74_151** (8-to-1 MUX with enable input) from the **SC Symbol** window. D0 through D7 are the data inputs. S0, S1 and S2 are the select inputs. EN is the enable input.

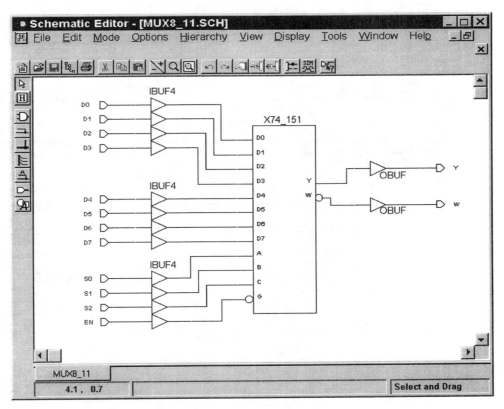

Figure 10.3 Schematic diagram for project **mux8_1**.

135

3) Perform a simulation for this design. Note that EN must be held LOW to enable the MUX to function normally. We will sample a few results here to show the characteristics of the 8-to-1 MUX.

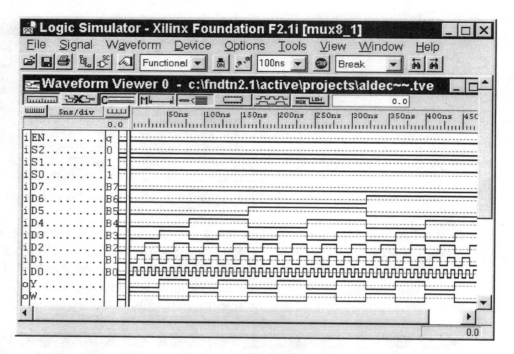

(a) EN = 0, S2S1S0 = 011, Y duplicates D3, $\overline{W}$ = Y.

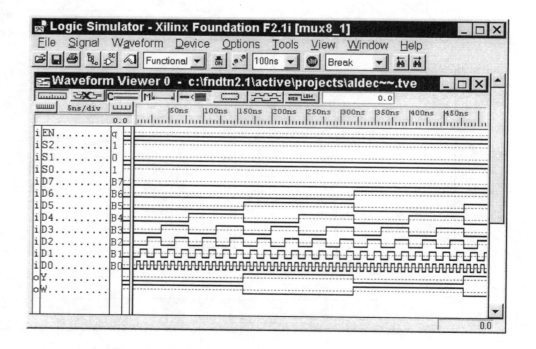

(b) EN = 0, S2S1S0 = 101, Y duplicates D5, $\overline{W}$ = Y.

Figure 10.4 Simulation result for the project *mux8*.

4)   Continue the simulation. Let EN = 1 and assign arbitrary values to S2, S1, and S0. Do the output waveforms for Y and W look similar? How? Why?

5)   Try other input combinations on the select lines and comment on what the MUX does.

Checked by _____ Date _____

Section II. The 1-of-8 DEMUX

1)   Create a new project called ***demux8*** in schematic mode. Your **New Project** window should look like Figure 10.5. Click on the **OK** button to go to the **Project Manager** window.

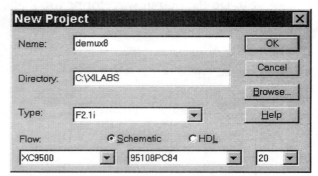

Figure 10.5 New Project window for ***demux8***.

2)   In the **Project Manager** window, click on the **Schematic Editor** button on the **Design Entry** tab to get into the **Schematic Editor** window. Create a schematic diagram similar to Figure 10.6 below. Note that we used **D3_8E** (1-of-8 decoder/demultiplexer with enable input) from the **SC Symbol** window. As the name indicates, DATA_INPUT is (what else) the data input. S0, S1, and S2 are the select inputs. O0 through O7 are the eight outputs.

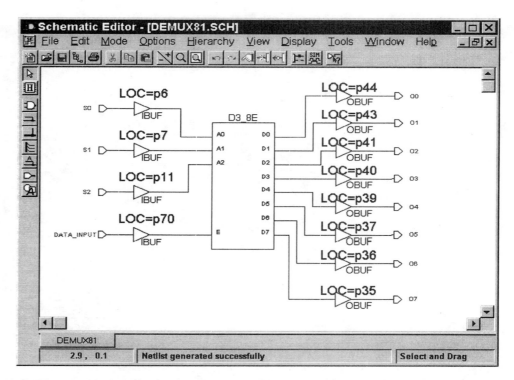

Figure 10.6 Schematic diagram with pin assignment for ***demux8***.

138

3) Make the pin assignment for your design. For the PLDT-1 board, toggle switch 8 is used for DATA_INPUT, and toggle switches 1, 2, and 3 for select inputs S0, S1, and S2, respectively. The eight outputs, O0 through O7, are displayed on LEDs 1 through 8, respectively.

4) Save the schematic design. Go through **Options → Create Netlist, Options → Integrity Test** and **Options → Export Netlist…** from the menu. Then, close the **Schematic Editor** window.

5) Implement this design and then download the bitstream onto your target board.

6) Test the design by using the SELECT and ENABLE input combinations shown in the following table. Note that S0 (assigned to toggle switch 1) is the LSB of the SELECT lines.

S2	S1	S0	DATA_INPUT	O0	O1	O2	O3	O4	O5	O6	O7
X	X	X	X								
X	X	X	X								
X	X	X	X								
0	0	0	0, 1								
0	0	1	0, 1								
0	1	0	0, 1								
0	1	1	0, 1								
1	0	0	0, 1								
1	0	1	0, 1								
1	1	0	0, 1								
1	1	1	0, 1								

Table 10.2 Function Table for the 1-of-8 Decoder/Demultiplexer

7) Comment on the function of the DEMUX.

Checked by _____ _____ Date _____

Section III.   The 1-of-8 DEMUX

Create the design shown in Figure 10.7 below. Implement the design and program it onto your target board. Do the following:

1)   Analyze the operation.

2)   List the function table (truth table).

3)   Give an example of where could we apply this circuit.

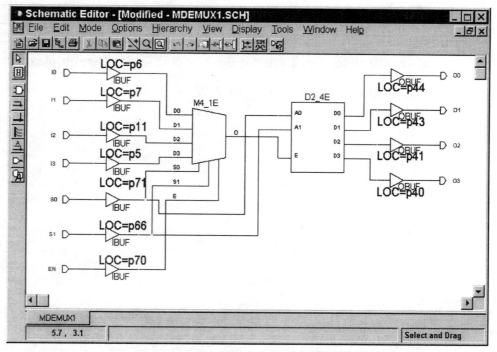

Figure 10.7 An application of MUX and DEMUX.

QUESTIONS:

1) Name two applications of MUXs.

2) Given a 4-to-16 decoder with an enable line, draw a block diagram showing how this decoder can be used as a 1-of-16 DEMUX.

3) Given two 8-to-1 MUXs with enable lines, how do you make a 16-to-1 MUX? Draw the block diagram.

4) Given a 4-to-1 MUX (M4_1E in the SC symbol list), how do you obtain the Boolean function:

$$O = \overline{S1}\ S0 + S1\ \overline{S0}$$

Show the external connection to the MUX.

NAME _____ **DATE** _____

CPLD EXPERIMENT 11:

Latches: S-R & D-Type

OBJECTIVES:

- Examine S-R, gated S-R, and gated D-type latches.
- Create the designs for the S-R, gated S-R, and gated D latches in schematic mode.
- Test the designs on the target board.

MATERIALS:

- Xilinx Foundation Software, student or professional edition V1.5 or higher.
- IBM or compatible computer with Pentium processor or equivalent, 64 M-byte RAM or more, and 3 G-byte or larger hard drive.
- PLDT-1 board by RSR Electronics Inc., XS95 board and XStend Board by XESS Corp., or a similar board with an XC95108 device.

DISCUSSION:

With this experiment, we start the discussion of sequential circuits. The main difference between combinational circuits and sequential circuits is that combinational circuits do not have memory elements. So the output of a combinatorial circuit depends only on the present inputs. But the output of a sequential circuit depends on the effects of prior inputs (the memory) as well as on the present inputs. Latches are a simple, but very important, class of memory elements.

The S-R NOR Latch

The S-R NOR latch has two inputs: S and R (SET and RESET) and two outputs: Q and not Q. The Q is the normal output and not Q is the complemented output. Any latch has two states: SET and RESET (CLEAR). When Q = 1, we say the latch is in the SET state. When Q = 0, the latch is in the RESET state. Figure 11.1 shows the construction of a NOR latch. (The notation S-C, SET & CLEAR, is sometimes used for S-R latches.)

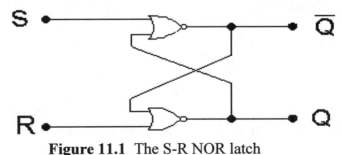

Figure 11.1 The S-R NOR latch

The truth table below (Table 11.1) describes the characteristics of this NOR latch.

Set (S)	Reset (R)		Output	State
0	0		No change	Latched
0	1		Q = 0	Cleared
1	0		Q = 1	Set
1	1		not Q = Q = 0	Invalid

Table 11.1 Truth Table for a NOR Latch

A NOR latch has active-high inputs. When both inputs are low (S,R = 0,0), the output can not change; it's "latched". Normally, S-R will go from 0,0 to a valid input (0,1 or 1,0) and then back to 0,0. When both inputs are high (1,1), both outputs are low, which is not valid since Q and not-Q should be opposites. As a result, when the inputs return to the latch state (S,R = 0,0), the outputs will be indeterminate (i.e., you can't predict if Q will be 0 or 1).

The Gated S-R Latch

In applications, we often need to have control over when a latch can change state. An enable line (EN) is used for the purpose. As shown in the Figure 11.2, two more gates are added to obtain the gated S-R latch. The gated S-R latch is also called the level-triggered S-R flip-flop (S-R FF) since Q can change only when EN "pulls the trigger".

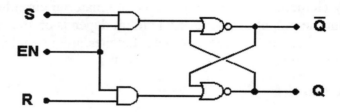

Figure 11.2 Diagram for the gated S-R latch.

The truth table in Table 11.2 shows how the EN input controls when the latch can respond to the S-R inputs.

EN	Set (S)	Reset (R)		Output
1	0	0		No change
1	0	1		Q = 0
1	1	0		Q = 1
1	1	1		Invalid, Q = not Q = 0
0	X	X		No change

Table 11.2 Truth Table for the Gated S-R Latch

It is seen that the function of the EN input is to enable/disable the inputs S and R.

The Gated D Latch

The gated D latch (D for data) can be built by adding an inverter before each of the two inputs on a gated S-R latch. A gated D latch is also called a level-triggered D flip-flop (D FF). Its diagram is shown in Figure 11.3.

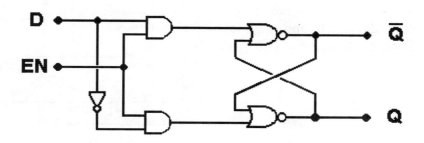

Figure 11.3 Diagram for the gated D latch.

By examining the following truth table, we can see that a level-triggered D FF has a simple operation. The output Q simply follows the data input D when the enable input is activated. Q is latched when the enable is low. We also see that the invalid state is gone.

EN	Data (D)		Output
1	0		Q = 0
1	1		Q = 1
0	X		No change

Table 11.3 Truth Table for the Gated D Latch

PROCEDURE:

Section I.   The NOR Latch and NAND Latch

1) Create a new project called *sr_latch* in schematic mode. Your **New Project** window should look like Figure 11.4. Click on the **OK** button to go to the **Project Manager** window.

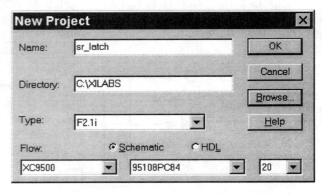

Figure 11.4 New Project window for *sr_latch*.

2) In the **Project Manager** window, create a schematic diagram similar to Figure 11.5 below. Note that we are creating two S-R latches: a NOR latch and a NAND latch.

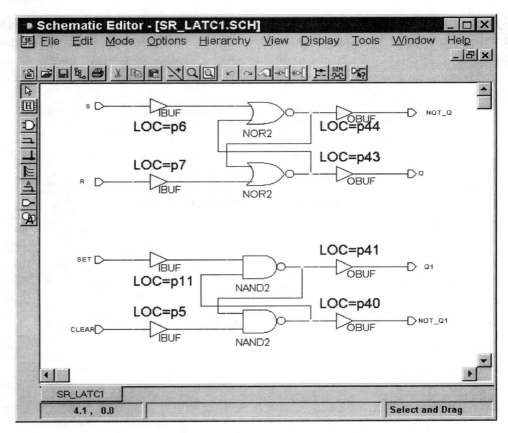

Figure 11.5 Schematic diagram with pin assignment for *sr_latch*.

146

3) Save the schematic design. Go through **Options → Create Netlist, Options →
Integrity Test** and **Options → Export Netlist...** from the menu. Then, close the
Schematic Editor window.

4) Implement this project and then download the bitstream onto your target board.

5) Determine from the schematic diagram which toggle switches and LEDs are used
for representing which inputs and outputs. Then, fill the output column in the
following table and then compare the operation of the NOR latch with Table 11.1.
Does it operate as you expected?

Preset (S)	Clear (R)		Output	State
0	0			
0	1			
1	0			
1	1			

Table 11.4 Experimental Results for the NOR Latch

6) Test the NAND latch on your target board and complete the following truth table.

Preset (S)	Clear (R)		Output	State
0	0			
0	1			
1	0			
1	1			

Table 11.5 Experimental Results for the NAND Latch

a) Are the inputs active-high or active-low?

b) Does the latch have an invalid state?

c) What are Q and not-Q when the outputs are invalid?

7) Compare the characteristics of the NOR latch with those of the NAND latch and
comment on the differences and similarities of these two latches.

Checked by _____ Date _____

147

Section II. The Gated S-R and D Latches

1) Create a new project called **g_latch** in schematic mode. Your **New Project** window should look like Figure 11.6. Click on the **OK** button to go to the **Project Manager** window.

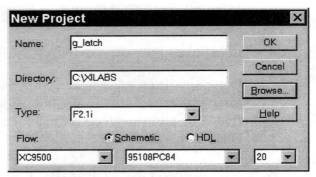

Figure 11.6 New Project window for **g_latch**.

2) In the **Project Manager** window, create a schematic diagram similar to Figure 11.7 below. Note that toggle switches 1, 2, and 3 are used for inputs S, R, and EN, respectively. LEDs 1 and 2 represent Q1 and not Q1. Toggle switches 7 and 8 provide inputs for D and GATE. LEDs 7 and 8 display Q2 and not Q2.

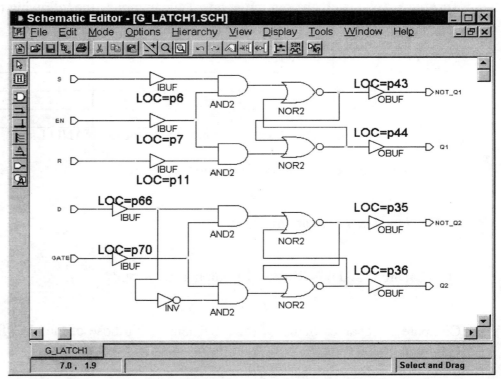

Figure 11.7 Schematic diagram with pin assignment for **g_latch**.

3) Save the schematic design. Go through **Options → Create Netlist, Options → Integrity Test** and **Options → Export Netlist...** from the menu. Then, close the **Schematic Editor** window.

148

4) Implement this design and then download the bitstream onto your target board.

5) Test the design by using all 1-0 combinations on the inputs. Fill in the output columns in the Table 11.6 and Table 11.7. Are the results the same as shown in tables 11.2 and 11.3 ?

EN	Set (S)	Reset (R)		Output
1	0	0		
1	0	1		
1	1	0		
1	1	1		
0	X	X		

Table 11.6 Experimental Results for the Gated S-R Latch

EN	Data (D)		Output
1	0		
1	1		
0	X		

Table 11.7 Experimental Results for the Gated D Latch

6) Describe the functions of the gated S-R and D latches in your own words.

Checked by _____ Date _____

QUESTIONS:

1) Draw the logic diagram for a gated S-R latch using only NAND gates.

2) How are the Gate or Enable inputs used in the gated latches?

3) If the enable input is to be active-low, what change(s) do you need to make to the gated D latch? Draw a diagram and show them.

4) How do you avoid the invalid state on an S-R latch?

CPLD EXPERIMENT 12:

Flip-Flops: J-K & D-Type

OBJECTIVES:

- Examine the characteristics of the JK FF and the D FF.
- Examine the difference between synchronous and asynchronous inputs.
- Examine the difference between level-triggering and edge-triggering.
- Study the characteristics of J-K and D flip-flops with asynchronous inputs.
- Create schematic designs for J-K and D FFs, including asynchronous inputs.
- Test the designs on the target board.

MATERIALS:

- Xilinx Foundation Software, student or professional edition V1.5 or higher.
- IBM or compatible computer with Pentium processor or equivalent, 64 M-byte RAM or more, and 3 G-byte or larger hard drive.
- PLDT-1 board by RSR Electronics Inc., XS95 board and XStend Board by XESS Corp., or a similar board with an XC95108 device.

DISCUSSION:

We have studied the level-triggered gated latch. In contrast to latches, most flip-flops (FFs) are edge-triggered. The symbols for both positive and negative edge-triggered FFs are shown in Figure 12.1 below. The control inputs are synchronous, meaning they will affect the outputs only when an "edge" occurs. A positive edge-triggered FF responds to its control inputs when triggered by a 0-to-1 (positive) transition on the clock input (CLK). Similarly, a negative edge-triggered FF can "look at" its control inputs only when a 1-to-0 (negative) transition occurs on CLK. Note the "bubble" on the CLK input of the FF in figure 12.1 (b). It indicates that the device is triggered by a falling edge.

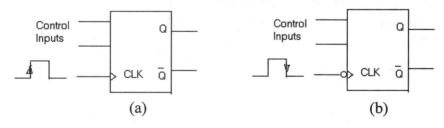

(a)   (b)

Figure 12.1 (a) Symbol for a positive edge-triggered FF.
(b) Symbol for a negative edge-triggered FF.

The Clocked J-K FF

Figure 12.2 below shows the symbol for a clocked J-K FF triggered by a negative edge. The truth table for this FF is listed in Table 12.1.

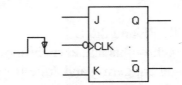

Figure 12.2 Symbol for negative edge-triggered J-K FF.

Inputs				Output		
J	**K**	**CLK**		**Q before CLK**	**Q after CLK**	**Description of Action**
0	0	↓		X	X	HOLD: no change in Q
0	1	↓		X	0	RESET (or Clear) Q to 0
1	0	↓		X	1	SET Q to 1
1	1	↓		X	$\overline{X}$	TOGGLE: Change Q to the opposite state

Table 12.1 Truth Table for Negative Edge-Triggered J-K FF (X = 0 or 1)

J and K are synchronous inputs. When the FF is clocked by a falling edge, one of three things will happen: 1) the Q output will match the J input if J and K are in opposite states; 2) Q will not change state if J and K are both LOW; 3) Q toggles to the opposite state if both J and K are HIGH. When there is no triggering, nothing will happen. Note that a J-K FF does not have an invalid state. The truth table for a positive edge-triggered J-K FF is the same as above except that the clock input is filled by ↑ (rising edge).

The Clocked D FF

The D FF has only one synchronous input and a clock input. Figure 12.3 shows the symbol for the positive edge-triggered D FF. Table 12.2 illustrates how the outputs respond to the inputs.

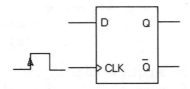

Figure 12.3 Symbol for Positive Edge-Triggered J-K FF

Inputs			Output		
D	CLK		Q before CLK	Q after CLK	Description of Action
0	↑		X	0	RESET (CLEAR) Q to 0
1	↑		X	1	SET Q to 1

Table 12.2 Truth table for Positive Edge-Triggered D FF (X = 0 or 1)

D (Data) is the synchronous input. The above table demonstrates that the output Q of a D FF follows D when triggered. The truth table for a negative edge-triggered D FF is the same as above except that the clock input is filled by ↓ (negative edge).

The J-K FF with Asynchronous Inputs

Unlike synchronous inputs, the asynchronous inputs $\bar{S}_D$ (DC SET) and $\bar{R}_D$ (DC RESET) do not depend on the clock. Moreover, they override the synchronous and clock inputs. The functions of the two asynchronous inputs are described in Table 12.3 below.

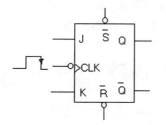

Figure 12.4 Symbol for J-K FF with asynchronous inputs.

Inputs						Output	
$\bar{S}$	$\bar{R}$	J	K	CLK		Q after CLK	Type of Operation
1	1	0	0	↓		Hold	Synchronous
1	1	0	1	↓		0	Synchronous
1	1	1	0	↓		1	Synchronous
1	1	1	1	↓		Toggle	Synchronous
0	1	X	X	X		1	Asynchronous
1	0	X	X	X		0	Asynchronous
0	0	X	X	X		Not used	Invalid

Table 12.3 Truth Table for J-K FF Shown in Figure 12.4 (X = 0 or 1)

The $\bar{S}$ and $\bar{R}$ are active-low asynchronous inputs. When they are both inactive (HIGH), the operation is synchronous and the J-K FF functions as if the asynchronous inputs do not exist. When one of asynchronous inputs is activated (LOW) the operation of the J-K FF is asynchronous and Q is forced to the state that the activated input (S or R) specifies.

153

PROCEDURE:

Section I. The Clocked D FF

1) Create a new project called *d_ff* in schematic mode. Your **New Project** window should look like Figure 12.5. Click on the **OK** button to go to the **Project Manager** window.

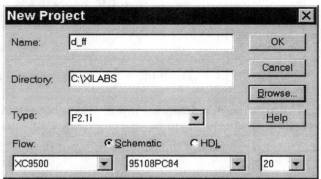

Figure 12.5 New Project window for *d_ff*.

2) In the **Project Manager** window, create a schematic diagram similar to Figure 12.6 below. We will examine the characteristics of the clocked D FF. We can get the clocked D FF (FD) from the **SC Symbol** list. The CLK signal is assigned to pin 10 on the PLDT-1 board, which is connected to the pulse switch.

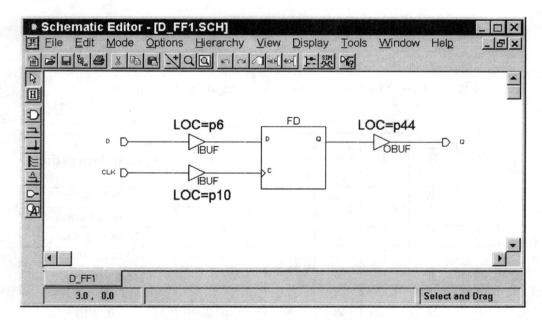

Figure 12.6 Schematic diagram with pin assignment for *d_ff*.

3) Save the schematic design. Go through **Options → Create Netlist, Options → Integrity Test** and **Options → Export Netlist...** from the menu. Then, close the **Schematic Editor** window.

4) Implement this project and then download the bitstream onto your target board.

5) Test the design on your target board. Complete the following table and verify the operation of the D FF with Table 12.2.

Inputs			Output	Description
D	CLK		Q After CLK	
0				
1				

Table 12.4 Experimental Results for the Clocked D FF

6) Compare the characteristics of the clocked D FF with the level-triggered D FF in CPLD Experiment 11 and comment on the difference and similarities of the two flip-flops.

Checked by _____ Date _____

Section II. The J-K FF with Asynchronous Preset and Clear Inputs

1) Create a new project called *jkff_sc* in schematic mode. Your **New Project** window should look like Figure 12.7. Click on the **OK** button to go to the **Project Manager** window.

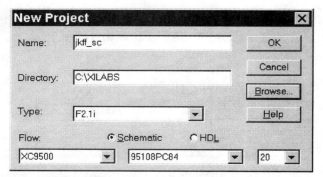

Figure 12.7 New Project window for *jkff_sc*.

2) In the **Project Manager** window, create a schematic diagram similar to Figure 12.8 below. Toggle switches 1, 2, 3, and 4 represent inputs J, K, SET, and CLEAR, respectively. The pulse switch supplies the CLK signal. The output Q is displayed on LED 1.

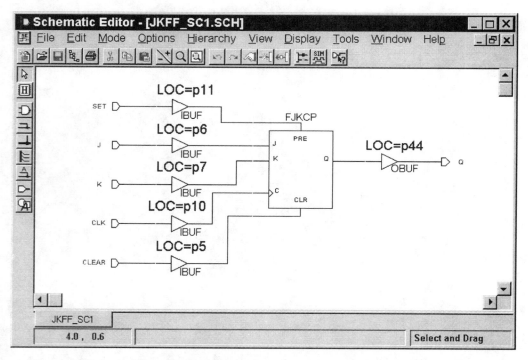

Figure 12.8 Schematic diagram with pin assignment for *jkff_sc*.

3) Save the schematic design. Go through **Options → Create Netlist, Options → Integrity Test** and **Options → Export Netlist...** from the menu. Then, close the **Schematic Editor** window.

4) Implement this design and then download the bitstream onto your target board.

156

5) Examine the characteristics of this J-K FF and complete the following table.

Inputs					Output	
PRE	CLR	J	K	CLK	Q after CLK	Type of Operation

Table 12.5 Experimental Results for the J-K FF with Asynchronous Preset and Clear

6) Answer the following:
   a) Are the asynchronous inputs PRE and CLR active-high or active-low?

   b) Is this J-K FF positive edge or negative edge triggered?

   c) Which type of inputs, synchronous or asynchronous, has higher priority?

   d) What does "toggle" mean? Is it a synchronous operation?

7) What is the difference between this J-K FF and the J-K FF shown in Figure 12.4?

Checked by _____ Date _____

157

QUESTIONS:

1) Draw the logic diagram for a positive edge-triggered D FF.

2) Name two applications of J-K and D FFs.

3) Are there other types of J-K FFs available in the **SC Symbol** list? Give two examples. Draw the symbols and write the truth tables.

4) We mentioned in the discussion that a FF has two outputs: Q and not-Q (sometimes called "Q-not" or "Q-bar"). Why doesn't the not-Q output appear on the J-K FF and D FF symbols in the **SC Symbol** list?

CPLD EXPERIMENT 13:
Asynchronous Counters

OBJECTIVES:

- Design asynchronous counters using both J-K and T type flip-flops.
- Design Modulo-N ripple counters in schematic mode.
- Test the counter designs on the target board.
- Use simulation to demonstrate the use of counters as frequency dividers.

MATERIALS:

- Xilinx Foundation Software, student or professional edition V1.5 or higher.
- IBM or compatible computer with Pentium processor or equivalent, 64 M-byte RAM or more, and 3 G-byte or larger hard drive.
- PLDT-1 board by RSR Electronics Inc., XS95 board and XStend Board by XESS Corp., or a similar board with an XC95108 device.

DISCUSSION:

Counters are an application of flip-flops (FFs). They are used in the many applications that require events to be counted or time intervals to be measured. There are two types of counters: synchronous and asynchronous. In this experiment, we will study how to design an asynchronous counter (also called a "ripple" counter). Regardless of the count sequence, if a counter has N counts, we call it a MOD-N counter. For example, if a counter counts from 2 to 6, it is a MOD-5 counter. If a counter counts in ascending order (0,1,2...), it's called an up-counter. If a counter counts in descending order (...2,1,0), it's called a down-counter. The following diagram illustrates an asynchronous MOD-8 up-counter.

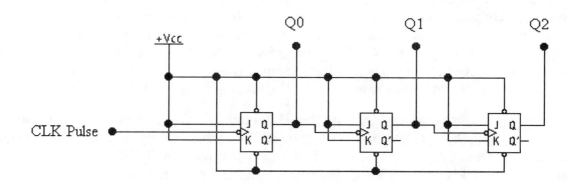

Figure 13.1 Logic diagram for a ripple MOD-8 up-counter.

The asynchronous counter shown in Figure 13.1 above uses three J-K flip-flops with asynchronous SET and CLEAR active-low inputs. Outputs are taken from Q0 (the LSB), Q1 and Q2 (the MSB). The clock signal is connected only to the first stage. The input of each successive stage is connected to the Q output of the preceding stage. The synchronous inputs J and K of each FF are tied HIGH so that its Q output will toggle when a 1-to-0 transition occurs on its CLK input. The asynchronous inputs SET and CLEAR are not used in this design, so they are connected to Vcc. It is important to note that the term "asynchronous counter" is not due to asynchronous inputs on the FFs, but to the fact that each FF (except the first) does not change state until it is triggered by the previous FF. Each clock pulse causes change to "ripple" from one FF to the next.

The following figure shows the timing waveforms of the MOD-8 ripple counter.

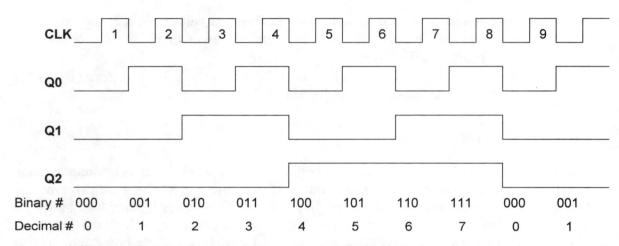

Figure 13.2 Timing waveforms for the MOD-8 ripple counter.

The outputs of the FFs form a 3-bit binary number. Assume that the counter has initial condition Q2Q1Q0 = 000. Since the FFs in the MOD-8 counter are negative edge-triggered and all the Js and Ks are connected to 1, Q0 toggles when it sees a 1-to-0 transition on the clock input, Q1 toggles when Q0 changes from 1 to 0, and Q2 toggles when Q1 changes from 1 to 0. So the binary number increases from 000 (0) to 111 (7_{10}) after seven clock pulses. After 111, the count "wraps around" to 000 and the sequence repeats itself as long as the CLK input receives pulses. For a counter with N FFs, the maximum number of counts is 2^N. For example, if we use 4 FFs to make a counter, the maximum number of counts would be 16 (2^4), ranging from 0000 (0) to 1111 (15).

Comparing the waveform of Q0 with the CLK input, we find that FF-1 (Q0) divides the CLK frequency by 2 (2^1), FF-2 (Q1) divides CLK frequency by 4 (2^2), and FF-3 (Q2) divides the CLK frequency by 8 (2^3). If the CLK frequency is, say, 8 kHz, the frequency at Q2 will be 8 kHz ÷ 8 = 1 kHz. Therefore, a MOD-8 counter is also called divide-by-8 counter.

In this experiment, we will design a few different ripple counters and examine how they work using the target board and simulation tools.

PROCEDURE:

Section I.   Design A MOD-2^N Ripple Counter

1)   Create a new project called **acount16** in schematic mode. Your **New Project** window should look like Figure 13.3. Click on the **OK** button to go to the **Project Manager** window.

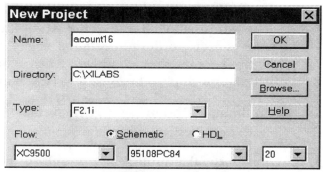

Figure 13.3 New Project window for **acount16**.

2)   In the **Project Manager** window, create a schematic diagram similar to Figure 13.4 below. This schematic is very similar to the ripple MOD-8 counter we examined in the discussion section except for the additional J-K FF. The ground (GND) and +5V (VCC) are available the **SC Symbol** list. For the PLDT-1 board, the CLK input is provided by a pulse switch (pin 10). Toggle switch 1 (Pin 6) is used to reset the counter. LEDs 1 to 4 are used to display the counter outputs Q3, Q2, Q1, and Q0, respectively.

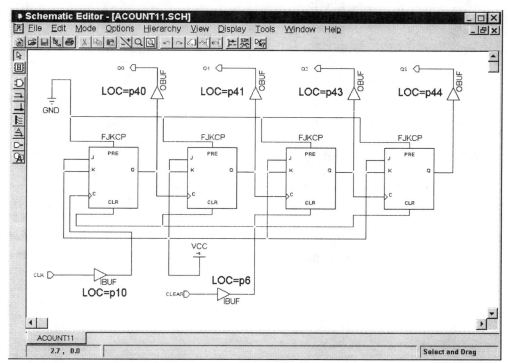

Figure 13.4 Schematic diagram with pin assignment for **acount16**.

Note that the OBUF symbols are rotated. Any symbol in the **Schematic Editor** window can be rotated clockwise (to point down) or counter-clockwise (to point up) by 90°. A symbol can also be "reversed" in two ways: a 180° rotation, or a mirror-image flip. To rotate a symbol, double click on the left button of the mouse to get the **Symbol Properties** window, and inside that window, click on the **Attributes** button. The **Symbol Attributes** window will appear. In the **Rotation** pane, select the desired rotation angle, click on the **OK** button and return to **Symbol Properties** window. Then, click on the **OK** button in the **Symbol Properties** window to go back to **Schematic Editor** window.

3)   Go through **Options → Create Netlist**, **Options → Integrity Test** and **Options → Export Netlist...** from the menu. Then, close the **Schematic Editor** window.

4)   Implement this project and download the bitstream onto your target board. On your target board, reset the counter by flipping the toggle switch to HIGH and then back to LOW. Observe the count sequence and record it on the following table as you push the pulse switch once for each clock pulse.

Clock Pulse	Output Q3 Q2 Q1 Q0	Base 10	Clock Pulse	Output Q3 Q2 Q1 Q0	Base 10
1st			9th		
2nd			10th		
3rd			11th		
4th			12th		
5th			13th		
6th			14th		
7th			15th		
8th			16th		

Table 13.1 Experimental Results for the MOD-16 Ripple Counter

5)   Determine the following:

a)   Is it an up-counter or a down-counter?

b)   Based on the count sequence you observed in step 4, how does this counter behave compared to the one shown in Figure 13.1?

c)   Why are the PRE inputs tied to GND?

d)   If we don't want a reset function, to where should we connect the CLR inputs of the flip-flops?

e)   What modification can you make so that this counter counts up? If there are two or more ways, list them all.

Checked by _____ Date _____

6) Simulate the counter. Your results should look similar to Figure 13.5 below.

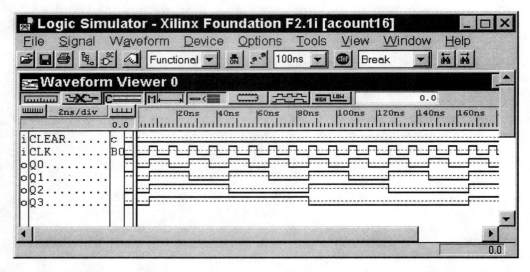

Figure 13.5 Timing waveforms for the asynchronous MOD-16 down-counter.

7) The CLEAR input is fed by switch "c" on the **Stimulator Selection** window. If it is connected to HIGH, what will you see on the **Waveform Viewer** window?

8) Read the time period of the CLK from the **Waveform Viewer** window and use it to calculate the frequency of the CLK.

9) Calculate the frequency at output Q3.

10) What is the "divide-by" number of this counter?

Checked by _____ Date _____

Section II. Design a MOD-N Ripple Counter with Arbitrary N

We have seen how to design MOD-N ripple-counters when N is a power of 2 (2^P). Sometimes we need counters with other modulo numbers, such as MOD-5 or MOD-10. We will design such a counter in this section of the experiment.

1)   Create a new project called **acount6** in schematic mode. Your **New Project** window should look like Figure 13.6. Click on the **OK** button to go to the **Project Manager** window.

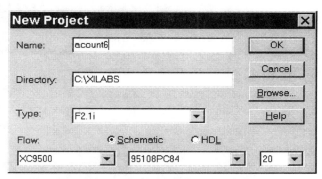

Figure 13.6 New Project window for *acount6*.

2)   In the **Project Manager** window, create a schematic diagram similar to Figure 13.7 below. The pulse switch on the PLDT-1 board supplies the CLK signal. The output Q2, Q1, and Q0 are displayed on LEDs 1, 2, and 3, respectively.

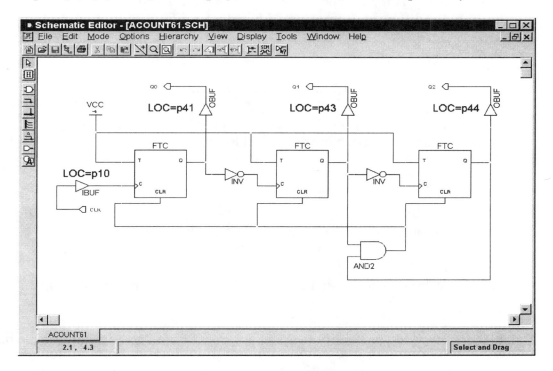

Figure 13.7 Schematic diagram with pin assignment for *acount6*.

3)   Go through **Options → Create Netlist, Options → Integrity Test** and **Options → Export Netlist...** from the menu. Then, close the **Schematic Editor** window.

4)   Implement this design and then download the bitstream onto your target board.

5)   Press the pulse switch and record the count sequence of this counter in the following table.

Clock Pulse	Output Q3 Q2 Q1 Q0	Base-10 Value
1st		
2nd		
3rd		
4th		
5th		
6th		
7th		
8th		

Table 13.2   Count Sequence for the Counter Design *acount6*

6)   Determine the following:

a)   What is the function of a T FF when the T input is tied to Vcc? What is the difference between a T FF and a J-K FF?

b)   What is the MOD number of this counter?

c)   What are the two inverters used for in the design?

d)   What role does the 2-input AND gate play?

e)   What will be the frequency at Q2 if the CLK frequency is 24 kHz?

f)   How do you modify the design to count from 7 to 3? Show the diagram.

Checked by _____   Date _____

166

QUESTIONS:

1)   How are asynchronous counters defined?

2)   Show how to make a T FF by making necessary external connections to a J-K FF.

3)   Draw a diagram of an asynchronous counter that counts from 6 to 1 using T FFs.

4)   Can you name a disadvantage of ripple counters?

NAME _____ **DATE** _____

CPLD EXPERIMENT 14:
TTL-Equivalent Library Counters

OBJECTIVES:

- Examine a few of the Xilinx Foundation library counters.
- Use library counters to design a divide-by-150 counter.

MATERIALS:

- Xilinx Foundation Software, student or professional edition V1.5 or higher.
- IBM or compatible computer with Pentium processor or equivalent, 64 M-byte RAM or more, and 3 G-byte or larger hard drive.
- PLDT-1 board by RSR Electronics Inc., XS95 board and XStend Board by XESS Corp., or a similar board with an XC95108 device.
- Function Generator.
- Oscilloscope.

DISCUSSION:

In CPLD experiment 13, we studied how to design asynchronous counters using J-K and T FFs. Since many designers are familiar with the performance of 7400 series digital integrated circuits, the **SC Symbol** list contains predefined counters modules that are equivalent to TTL ICs such as the 74160, 74161, 74162, and 74163. They are like predefined macros, and we can build a MOD-N counter quickly by using them.

PROCEDURE:

Section I. The 74160 BCD Counter

The BCD, or decade, counter is a MOD-10 counter. Since most numeric input/output devices use decimal numbers, it's convenient to have counters that operate in base-10.

1) Create a new project called **count10** in schematic mode. Your **New Project** window should look like Figure 14.1. Click on the **OK** button to go to the **Project Manager** window.

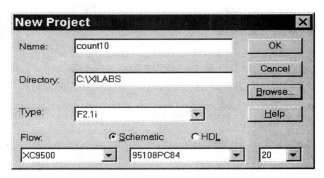

Figure 14.1 New Project window for **count10**.

2) In the **Project Manager** window, create a schematic diagram similar to Figure 14.2 below. Since we want the output on the 7-segment display, we'll use the ***deco7g*** macro (dec_7seg project) that we created in CPLD experiment 9. To add this symbol to our current library, choose **Tools → Project Libraries...** from the menu in the **Schematic Editor** window. In the **Project Libraries** window, browse through the **Attached Libraries** list on the left and select the file dec_7seg. Then click on the **Add** button. The dec_7seg project name should appear in the **Project Libraries** list on the right of this window. Click on the **Close** button to go back to the **Schematic Editor** window. Verify that ***deco7g*** is now included at the top of the **SC Symbol** list.

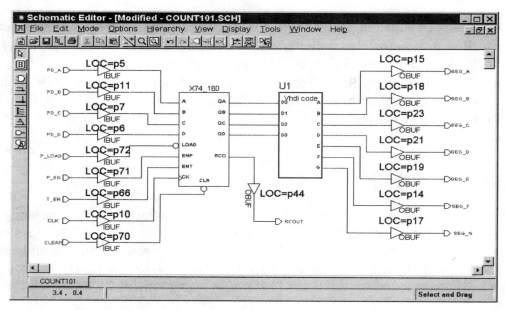

Figure 14.2 Schematic diagram for project ***count10***.

Toggle switches 1 through 4 represent parallel data PD_D, PD_C, PD_B, and PD_A, respectively. Toggle switches 5 through 8 represent P_LOAD (parallel load), P_EN (parallel enable), T_EN (trickle enable), and CLEAR inputs respectively. LED 1 is used to display the output RCOUT. The pulse switch is connected to the CLK input. On the target board, the 7-segment display is pre-wired: segments a, b, c, d, e, f, and g are connected to pins 15, 18, 23, 21, 19, 14, and 17, respectively.

3) Go through **Options → Create Netlist**, **Options → Integrity Test** and **Options → Export Netlist...** from the menu. Then, close the **Schematic Editor** window.

4) Implement the design and then download the bitstream onto your target board.

5) Examine the functions of the 74160 module and complete the following table.

Inputs

CK	$\overline{\text{CLR}}$	$\overline{\text{LOAD}}$	ENP	ENT		Function
X	0	X	X	X		
↑	1	0	X	X		
↑	1	1	0	0		
↑	1	1	0	1		
↑	1	1	1	0		
↑	1	1	1	1		

Table 14.1 Truth Table for the 74160 Counter IC ($X = 0$ or 1)

6) Determine the following for the 74160:

a) When will output RCO (LED 1) go HIGH? What could RCO be used for?

b) What are ENT and ENP used for?

c) Is CLEAR a synchronous or an asynchronous operation? Explain.

d) Is LOAD a synchronous or an asynchronous operation? Explain.

Checked by _____ Date _____

171

Section II.   The 74163 4-Bit Counter

1)   Create a new project called ***count12*** in schematic mode. Your New Project window should look like Figure 14.1. Click on the **OK** button to go back to the **Project Manager** window.

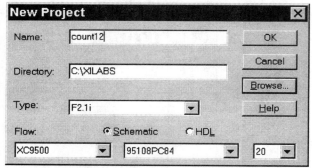

Figure 14.3 New Project window for ***count12***.

2)   In the **Project Manager** window, create a schematic similar to Figure 14.4 below.

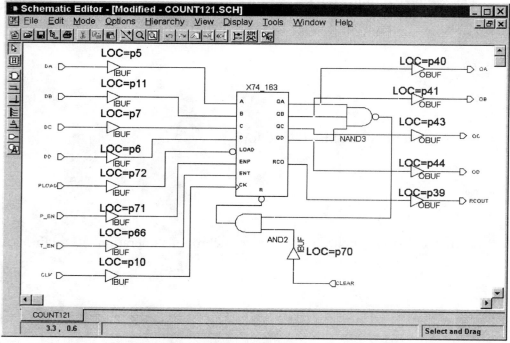

Figure 14.4 Schematic diagram for project ***count12***.

Toggle switches 1 through 4 generate parallel data DD, DC, DB, and DA, respectively.   Toggle switches 5 through 8 are PLOAD (parallel load), P_EN (parallel enable), T_EN (trickle enable), and RESET respectively. The pulse switch is connected to the CLK input. LEDs 1 to 4 display outputs OD, OC, OB, and OA, respectively. LED 5 is used to indicate output RCO.

3)   Go through **Options → Create Netlist, Options → Integrity Test** and **Options → Export Netlist...** from the menu. Then, close the **Schematic Editor** window.

4)   Implement this design and then download the bitstream onto your target board.

5)   Examine the functions of the 74163 module and complete the following table.

Inputs

CK	$\overline{R}$	$\overline{LOAD}$	ENP	ENT		Function
↑	0	X	X	X		
↑	1	0	X	X		
↑	1	1	0	0		
↑	1	1	0	1		
↑	1	1	1	0		
↑	1	1	1	1		

Table 14.2 Truth Table for the 74163 Counter

6)   Determine the following for the 74160:

a) When will output RCO (LED 1) go HIGH?

b) Is the RESET input synchronous or asynchronous? Explain.

c) Explain how NAND3 and AND2 are used to make the counter MOD-12.

d) If RESET was not synchronous, how would you configure the 74163 to make a MOD-12 counter? Draw a diagram to show how.

Checked by _____ Date _____

173

Section III. Design a Divide-by-150 Using Counter Modules

In some applications, it is necessary to slow down the clock rate. For example, in a PC system the clock rate for the CPU is hundreds of MHz, which has to be scaled down for the slower RAM and ROM devices. This can be done by using frequency dividers.

In this part of the experiment, you need to determine the following:

1) Which modules can be used to construct a divide-by-150 counter.

2) How should the modules be interconnected.

3) Where is the input signal and where is the output.

4) Which pins on the target board to use for the clock input from the function generator and for the output to the oscilloscope.

Once you have answered the above questions, perform the following:

5) Draw the schematic, with pin assignments, on the Schematic Editor for the divide-by-150 counter.

6) Implement and download the design to the target board.

7) Test the design on the target board. Use the TTL output of your function generator and adjust the frequency to 15 KHz. Use your scope to observe the output of your design and verify that the frequency is 100 Hz.

8) Draw the block diagram of your design below. You only need to show the interconnection between the modules, as well as the input and output.

Checked by _____ Date _____

QUESTIONS:

1)   What is the difference between synchronous CLEAR and asynchronous RESET?

2)   Draw the block diagram for a 2-digit BCD counter using counter ICs.

3)   Draw the block diagram for a divide-by-120 counter using counter ICs. Show the interconnections between the ICs, the clock input, and the divided-down output.

NAME _____ **DATE** _____

CPLD EXPERIMENT 15:
Sequential Design in HDL: A Synchronous Counter

OBJECTIVES:

- Examine the design of a universal synchronous counter in VHDL and ABEL.
- Gain experience with sequential circuit design in VHDL and ABEL

MATERIALS:

- Xilinx Foundation Software, student or professional edition V1.5 or higher.
- IBM or compatible computer with Pentium processor or equivalent, 64 M-byte RAM or more, and 3 G-byte or larger hard drive.
- PLDT-1 board by RSR Electronics Inc., XS95 board and XStend Board by XESS Corp., or a similar board with an XC95108 device.

DISCUSSION:

In a previous experiment, we designed a decoder circuit in VHDL. Decoders belong to the combinatorial class of circuits. In this experiment, we will show how to design sequential circuits in HDL mode. In particular, we will demonstrate the design of a 4-bit universal synchronous counter in both VHDL and ABEL.

The 4-bit universal synchronous counter has the following functions:

CLK	LOAD	RESET	CE	DIR		Function
X	X	1	X	X		Asynchronous Reset
↑	1	0	X	X		Synchronous Load
X	X	X	0	X		Hold
↑	0	0	1	1		Count Up
↑	0	0	1	0		Count Down

Table 15.1 Function Table for the Universal Synchronous Counter

Here CLK is the clock pulse, LOAD is the load enable, RESET is the reset input, CE is the count enable, and DIR is used to select count up or count down operation. Note that RESET is an asynchronous operation, while load and count are synchronous operations.

Section I.   Universal Synchronous Counter in VHDL

1) Create a new project called *u4count* in HDL mode. Your **New Project** window should look like Figure 15.1. Click on the **OK** button to go to the **Project Manager** window.

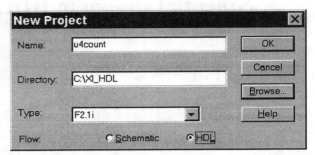

Figure 15.1 New Project window for *u4count*.

2) Click on the **HDL Editor** button on the **Design Entry** tab. In the **HDL Editor** selection window, select the **Use HDL Design Wizard** radio button and click on the **OK** button to go to the **Design Wizard** window. In the **Design Wizard** window, click on the **Next** button. Then, select the VHDL radio button and click on the **Next** button in the **Design Wizard – Language** window to get to the **Design Wizard – Name** window. In this window, type the name *u4cnt* for the *u4cnt.vhd* file and click on the **Next** button to go to the **Design Wizard – Ports** window.

3) In the **Design Wizard – Ports** window, define the nine inputs: D[3:0], CLK, RESET, CE, LOAD, and DIR. Also, define the four outputs: Q[3:0]. When this is done, click on the **Finish** button to go to the **HDL Editor** window.

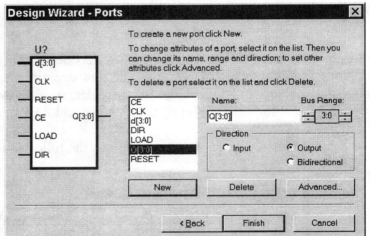

Figure 15.2 Input and output ports for *u4cnt*.

4) The skeleton of the *u4cnt.vhd* file is generated and shown in the **HDL Editor** window. As shown in Figure 15.3, the control and clock inputs are defined as STD_LOGIC (standard logic) type. This type declaration specifies the values that may appear on the input and the outputs as well as the operations that may be applied to the signals. The values include 0, 1, X (unknown), and U (uninitalized). The data inputs and outputs are defined as STD_LOGIC_VECTOR type. Since we have the elements in descending order, the keyword "downto" is shown. If we used the data inputs as D[0:3] in the **Design Wizard – Ports** window, it will appear in the **HDL Editor** window as D: in STD_LOGIC_VECTOR (0 to 3).

```
u4cnt - HDL Editor
File  Edit  Search  View  Synthesis  Project  Tools  Help

1   library IEEE;
2   use IEEE.std_logic_1164.all;
3
4   entity u4cnt is
5       port (
6           D: in STD_LOGIC_VECTOR (3 downto 0);
7           CLK: in STD_LOGIC;
8           RESET: in STD_LOGIC;
9           CE: in STD_LOGIC;
10          LOAD: in STD_LOGIC;
11          DIR: in STD_LOGIC;
12          Q: out STD_LOGIC_VECTOR (3 downto 0)
13      );
14  end u4cnt;
15
16  architecture u4cnt_arch of u4cnt is
17  begin
18      -- <<enter your statements here>>
19  end u4cnt_arch;
20

Ready                                    Ln 6, Col 10   VHDL      CAP NUM
```

Figure 15.3 Skeleton of *u4cnt.vhd*.

5)   We need to add the program body for the counter design. Since it is not our purpose to write a complete VHDL tutorial in this experiment, we will utilize a VHDL template available in the "Language Assistant" and tailor it to fit our need. Choose **Tools → Language Assistant** from the menu in the **HDL Editor** window to get into the **Language Assistant –VHDL** window. In this window, click on the "+" next to **Synthesis Templates** and highlight the item **Synchronous Counter**. The VHDL codes are displayed in the right pane as shown in Figure 15.4. Make sure the cursor in your **HDL Editor** window is positioned below the line saying **<<Enter your statements here>>** before you copy this template to

your *u4cnt.vhl* file. Then click on the **Use** button to copy the template and the "X" button on the upper right corner to close the window. The template for the synchronous counter will now appear in your **HDL Editor** window as shown in Figure 15.5.

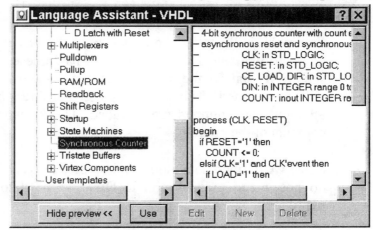

Figure 15.4 Template for the synchronous counter.

179

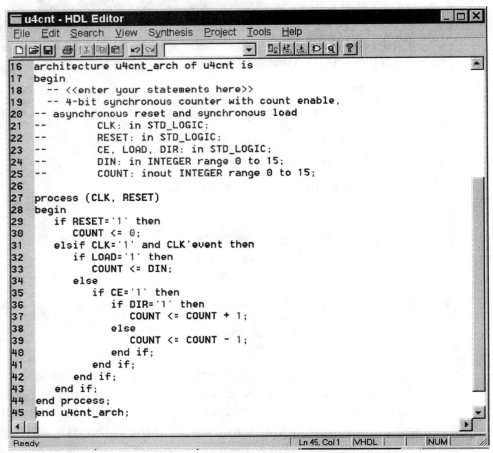

```
 u4cnt - HDL Editor                                        _ □ X
 File  Edit  Search  View  Synthesis  Project  Tools  Help
16  architecture u4cnt_arch of u4cnt is
17  begin
18    -- <<enter your statements here>>
19    -- 4-bit synchronous counter with count enable,
20  -- asynchronous reset and synchronous load
21  --        CLK: in STD_LOGIC;
22  --        RESET: in STD_LOGIC;
23  --        CE, LOAD, DIR: in STD_LOGIC;
24  --        DIN: in INTEGER range 0 to 15;
25  --        COUNT: inout INTEGER range 0 to 15;
26
27  process (CLK, RESET)
28  begin
29    if RESET='1' then
30      COUNT <= 0;
31    elsif CLK='1' and CLK'event then
32      if LOAD='1' then
33        COUNT <= DIN;
34      else
35        if CE='1' then
36          if DIR='1' then
37            COUNT <= COUNT + 1;
38          else
39            COUNT <= COUNT - 1;
40          end if;
41        end if;
42      end if;
43    end if;
44  end process;
45  end u4cnt_arch;

Ready                              Ln 45, Col 1   VHDL        NUM
```

Figure 15.5 VHDL template for the synchronous counter added to *u4cnt.vhd*.

The comments indicate what this template is designed for and the names for the input and output variables. Notice that we purposefully used some of the same names when we assigned the I/O ports in step 3.

Figure 15.5 illustrates the structure of the basic process, which begins with the keyword **process**. The two signals in parentheses, CLK and RESET, make up the **sensitivity list** for the process. Whenever one of the signals changes, the process will be executed. The sensitivity list is different from the parameter list in which all inputs and outputs are included. The keywords **begin** and **end** enclose the contents of the process.

Since we defined data inputs and outputs as STD_LOGIC_VECTOR type, but we do not intend to use variable DIN, we need to modify the statements in the template.

6)   The modified VHDL program for the universal synchronous counter is shown in Figure 15.6(a)(b). Note that you need to press <Enter> after the last line of the VHDL code.

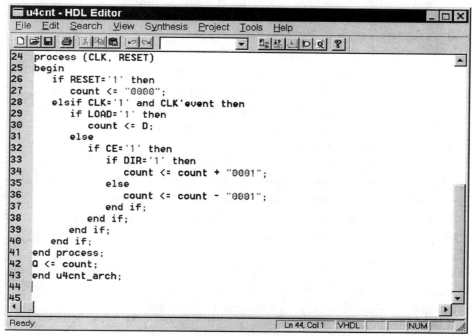

(a) Universal synchronous counter design in VHDL, lines 1 to 23.

(b) Universal synchronous counter design in VHDL, lines 24 to 43.

Figure 15.6

On line 3, the **std_logic_unsigned.all** package from the IEEE library is added to the program. This package defines unsigned number operations on type **std_logic**. Without this package, the operations indicated on lines 34 and 36 are illegal. The comments are added in the **entity** section so that we can omit the comments on lines 21 to 25 in Figure 15.5. On line 19, we declared the *count* to be a **signal** that is type **STD_LOGIC_VECTOR (3 downto 0)**. This means the **count** vector has four elements that are internal signals; they are evaluated at some future point in time and are not connected to any ports. We must make this declaration in order to implement the up-count and down-count shown on lines 34 and 36.

In VHDL, a positive-going-transition of the clock pulse is represented by the line CLK = '1' and 'CLK'event. Within the body of the **process** there are a few VHDL conditional structures, such as **if-then-else,** that are similar to those in C language. We changed the template on lines 27, 34 and 36 to adapt the variable type that we used. Since we did not declare the DIN signal as required by the template, we changed the statement on line 30 to assign vector **D** to **count** when there is a positive edge-triggered CLK and LOAD is enabled. The statement in line 42 is used to connect the vector **count** to output vector **Q**.

7) To check if there is any syntax error, choose **Synthesis** → **Check Syntax** from the menu in the **HDL Editor** window.

8) Choose **Project** → **Create Macro** from the menu. In the **Set initial target** window, make sure *XC9500*, *95108PC84,* and *20* are in the editor boxes **Family**, **Part**, and **Speed**, respectively. Click on the **OK** button. If everything is OK, you will see the pop up window indicating the symbol was created successfully. Click on the **OK** button.

9) The macro *u4cnt* is now created. Upon returning to your **Project Manager** window, go back to the **HDL Editor** window and choose **Project** → **Add to Project** from the menu. This is to include the macro in the project *u4count* so that we can perform a simulation to check the correctness of the design. Exit to the **Project Manager** window.

10) Synthesize the macro by clicking on the **Synthesis** button. When the contents of the editor boxes in the **Synthesis/Implementation settings** window are consistent with the design, click on the **Run** button. Both the **Design Entry** and **Synthesis** buttons now show green checkmarks.

11) Run a simulation to test the design. Figure 15.7(a)(b) display two functions of the universal synchronous counter design. Refer to Table 15.1 and verify the functions LOAD, RESET, HOLD, COUNT UP and COUNT DOWN using simulation tools. Toggles switches e, d, l, and r are assigned to CE, DIR, LOAD, and RESET, respectively.

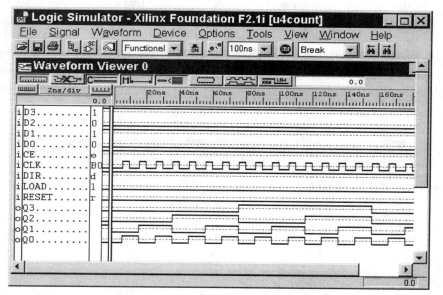

(a) When CE=1, LOAD=0, DIR=1, the counter counts up on positive edges on CLK.

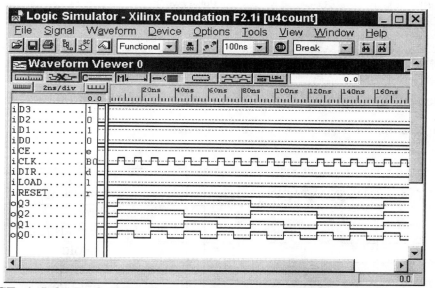

(b) When CE=1, LOAD=0, DIR=0, the counter counts down on positive edges on CLK.

Figure 15.7

Checked by _____ Date _____

183

12) To test the design on the PLDT-1 board, we need to add the pin assignment to the *u4count.ucf* file. To do this, right click on **u4count** (the top one in the **Files** folder) in the left pane of the **Project Manager** window. You should see a pull-down menu. In that menu, choose **Edit Constraints**. You now have the **Report Browser** window. In that window, enter the statements for the pin assignment as illustrated in Figure 15.8. Save this file and go back to the **Project Manager** window.

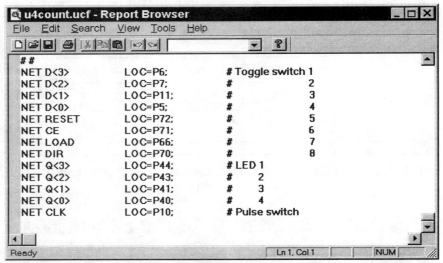

Figure 15.8 Pin assignment in *u4count.ucf* file for target board.

13) Implement and download the bitstream into your target board.

14) Complete the following truth table for the design.

Inputs

CLK	LOAD	RESET	CE	DIR		Function
						Asynchronous Reset
						Synchronous Load
						Hold
						Count Up
						Count Down

Table 15.2 Experimental Results for the Universal Synchronous Counter

15) Comment on the functions of the universal synchronous counter.

Checked by _____ Date _____

184

Section II. Universal Synchronous Counter in ABEL

1)   Create the *ablehdl* directory in your C drive.

2)   In Xilinx Foundation 2.1, to get into ABEL HDL mode, you must first start in schematic mode. Create a new project called *udcount* in schematic mode. Your **New Project** window should look like Figure 15.9. Click on the **OK** button to go to the **Project Manager** window.

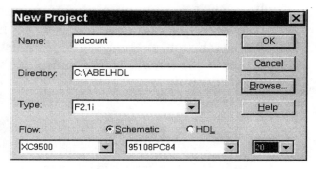

Figure 15.9 New Project window for *udcount*

3)   In the **Project Manager** window, click on the **HDL Editor** button on the **Design Entry** tab. The **HDL Editor** selection window will appear. Select the **Use HDL Design Wizard** radio button, and then click on the **OK** button.

4)   In the **Design Wizard** window, click on the **Next** button. In the **Design Wizard – Language** window, choose the **ABEL** option and click on the **Next** button. Type the name *udcnt* for the file *udcnt.abl* in the editor box of the **Design Wizard – Name** window, and then click on the **Next** button.

5)   You should now have the **Design Wizard – Port** window. In this window, you will specify the inputs and outputs of the universal synchronous counter. Repeat the instructions of Section I, step 3. Click on the **Finish** button to go to the **HDL Editor** window. You should see the skeleton of *udcnt.abl* shown in Figure 15.10.

```
udcnt.abl - HDL Editor
File  Edit  Search  View  Synthesis  Project  Tools  Help

1   module udcnt
2   Title 'udcnt'
3
4   Declarations
5
6   D3..D0 PIN;
7   D = [D3..D0];
8   CLK PIN;
9   RESET PIN;
10  CE PIN;
11  LOAD PIN;
12  DIR PIN;
13  Q3..Q0 PIN istype 'reg';
14  Q = [Q3..Q0];
15
16  "   <<add your declarations here>>
17
18  Equations
19
20  "   <<add your equations here>>
21
22  end udcnt
```

Figure 15.10 Skeleton of *udcnt.abl*.

185

6)   In the **HDL Editor** window, choose **Tools → Language Assistants** from the menu.   In the **Language Assistant – ABEL** window, click the "+" next to the **Synthesis Templates** and highlight **Synchronous Counter.** The ABEL code for the synchronous counter should appear in the right pane. Position the cursor below *<<Add your equations here>>* in the **HDL Editor** window and click on the **Use** button in the **Language Assistant – ABEL** window.   The template is added to the ***udcnt.abl*** file shown in the **HDL Editor** window as illustrated in Figure 15.11 below.

```
udcnt.abl - HDL Editor                                    _ □ ✕
File  Edit  Search  View  Synthesis  Project  Tools  Help

20 "   <<add your equations here>>
21 " 4-bit synchronous counter with count enable, asynchronous res
22 "         CLK            pin;
23 "         RESET          pin;
24 "         CE, LOAD, DIR  pin;
25 "         DIN3..DIN0     pin;
26 "         COUNT3..COUNT0 pin istype 'reg';
27 "
28 "         DIN = [DIN3..DIN0];
29 "         COUNT = [COUNT3..COUNT0];
30
31 equations
32
33 COUNT.CLK = CLK;
34 COUNT.ACLR = RESET;
35
36 when LOAD then COUNT := DIN;
37        else when !CE then COUNT := COUNT;
38                      else when DIR then COUNT := COUNT + 1;
39                                    else COUNT := COUNT - 1;
40
41 end udcnt

Ready                                  Ln 31, Col 1   ABEL        NUM
```

Figure 15.11   The template for a synchronous counter added to ***udcnt.abl.***

7)   Make changes in the ***udcnt.abl*** file as shown in Figure 15.12.   The program begins with the keyword **module** followed by the *macro name* and ends with the keyword **end** followed by the *macro name*.   The body of the program consists of the **declarations** section and the **equations** section.   The program structure is similar to VHDL, but the syntax is different.

In ABEL, .CLK and .ACLR are used to represent positive-edge clock trigger and asynchronous clear, respectively. So line 21 defines the *load, hold,* and *count up/down* operations as synchronous operations and line 22 indicates that the *reset* operation is asynchronous.   Note that we deleted the variables **count** and **DIN** from the template since ABEL allows us to update the variables on the I/O ports. Compared with VHDL files, ABEL files usually are shorter because ABEL is less rigid and less descriptive.

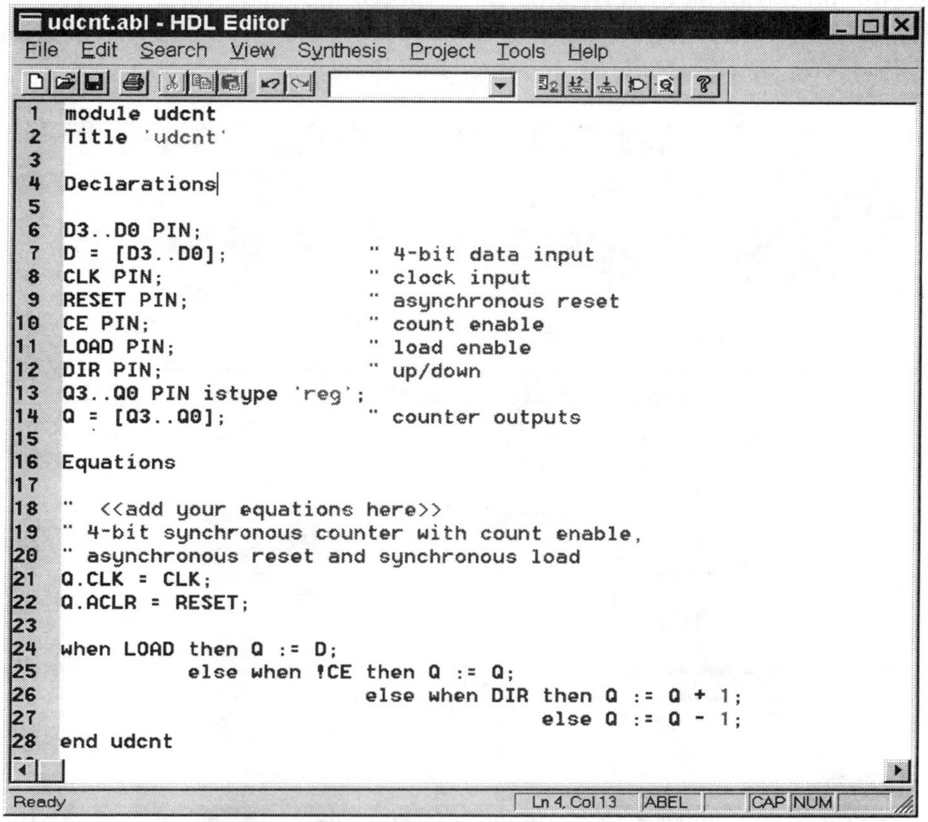

```
udcnt.abl - HDL Editor                                      _ □ ×
File  Edit  Search  View  Synthesis  Project  Tools  Help

1   module udcnt
2   Title 'udcnt'
3
4   Declarations
5
6   D3..D0 PIN;
7   D = [D3..D0];              " 4-bit data input
8   CLK PIN;                   " clock input
9   RESET PIN;                 " asynchronous reset
10  CE PIN;                    " count enable
11  LOAD PIN;                  " load enable
12  DIR PIN;                   " up/down
13  Q3..Q0 PIN istype 'reg';
14  Q = [Q3..Q0];             " counter outputs
15
16  Equations
17
18  "   <<add your equations here>>
19  " 4-bit synchronous counter with count enable,
20  " asynchronous reset and synchronous load
21  Q.CLK = CLK;
22  Q.ACLR = RESET;
23
24  when LOAD then Q := D;
25          else when !CE then Q := Q;
26                  else when DIR then Q := Q + 1;
27                          else Q := Q - 1;
28  end udcnt

Ready                               Ln 4, Col 13   ABEL      CAP NUM
```

Figure 15.12 ABEL code for the universal synchronous counter.

8) Synthesize the macro by clicking the **Synthesis** button. In the **HDL Editor** window, choose **Project → Create Macro** from the menu. Click on the **OK** button. The macro *udcnt* is created.

9) In the **HDL Editor** window, choose **Project → Add to Project** from the menu. That will include this macro in the project *udcount* so that we can perform a simulation. Save the *uncnt.abl* file and exit to the **Project Manager** window.

10) Synthesize the macro by clicking on the **Synthesis** button. Make sure the editor boxes in your **Synthesis/Implementation settings** are consistent with your project. Then, click on the **Run** button. Both the Design Entry and Synthesis buttons should show green check marks.

187

11) Use simulation tools to verify the functions of the universal synchronous counter. Figure 15.13(a)(b) below illustrate the count-up and count-down functions of the design. You should examine the other functions also.

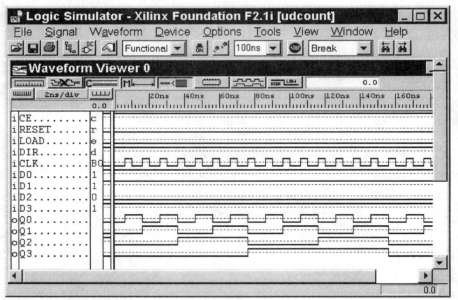

(a) When CE=1, LOAD=0, DIR=1, the counter counts up on positive edges on CLK.

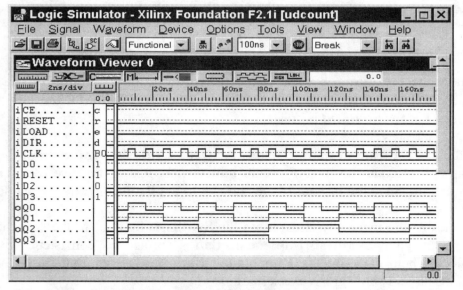

(b) When CE=1, LOAD=0, DIR=0, the counter counts down on positive-edges on CLK.

Figure 15.13

12) Test the design on the PLDT-1 board. Make pin assignment in the *udcnt.abl* file as shown in Figure 15.14. Then implement and download the bitstream. Does the design function the same as in your simulation?

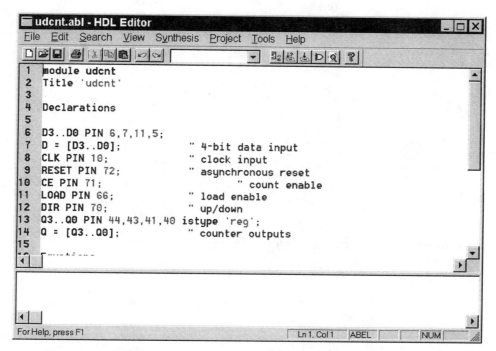

Figure 15.14 Pin assignment for project *udcount*.

Checked by _____ Date _____

QUESTIONS:

1) In Section I, where and how did you make the pin assignment? Write the statements.

2) How do you define a negative edge-triggered CLK?

CPLD EXPERIMENT 16:
Shift Registers & Ring Counters

OBJECTIVES:

- Examine a few of the register modules in the Xilinx symbol library.
- Construct a 16-bit register using two 8-bit register modules.
- Explore the operation of shift-registers, ring-counters, and Johnson counters.

MATERIALS:

- Xilinx Foundation Software, student or professional edition V1.5 or higher.
- IBM or compatible computer with Pentium processor or equivalent, 64 M-byte RAM or more, and 3 G-byte or larger hard drive.
- PLDT-1 board by RSR Electronics Inc., XS95 board and XStend Board by XESS Corp., or a similar board with an XC95108 device.
- Function Generator.
- Oscilloscope.

DISCUSSION:

Shift registers are widely used in data communication and computer systems to move and temporarily store data bits. They can handle parallel and serial movement of data bits, as well as conversions from parallel to serial and serial to parallel. In this experiment, we will examine a few TTL equivalent registers. By constructing and simulating the register designs, we will gain understanding on how the registers work.

PROCEDURE:

Section I. The 74194 4-Bit Bidirectional Universal Shift Register

The 74194 provides four operating modes. They are listed in Table 16.1 below, where inputs S0 and S1 are used to select the operating mode of the shift register.

S1	S0		Operating Mode
0	0		Hold
0	1		Shift right
1	0		Shift left
1	1		Parallel load

Table 16.1 Operating Modes of the 74194

In addition to the above four operating modes, the 74194 has an asynchronous RESET function (Master Clear). We will create the design and examine its various operations.

1) Create a new project called *reg_ic* in schematic mode. Your **New Project** window should look like Figure 16.1. Click on the **OK** button to go to the **Project Manager** window.

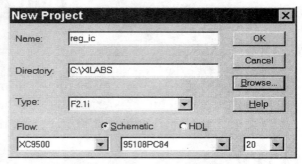

Figure 16.1 New Project window for *reg_ic*.

2) In the **Project Manager** window, create a schematic diagram similar to Figure 16.2 below.

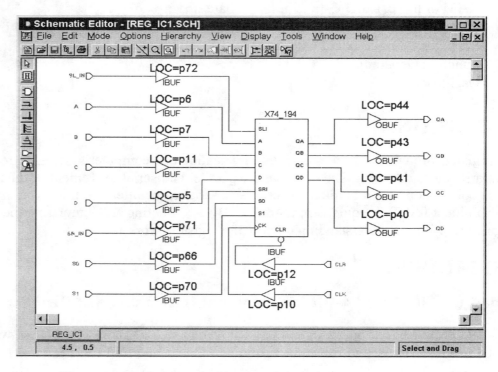

Figure 16.2 Schematic diagram for project *reg_ic*.

Here A, B, C, and D are the parallel data inputs. SL_IN and SR_IN represent shift-left data input and shift-right data input. S0 and S1 are the mode selection bits. CLK is the clock signal. CLR is the master clear. Note the pin assignment is also illustrated in Figure 16.2. For the PLDT-1 board, toggle switches 1, 2, 3, and 4 are used for the parallel data inputs A, B, C, and D, respectively. Toggle switches 5, 6, 7, and 8 represent SL_IN, SR_IN, S0 and S1, respectively. The clock signal CLK is provided by pulse switch and the asynchronous CLR is connected to pin 12 column of Tie Block 1. LEDs 1, 2, 3, and 4 are used to display parallel data outputs QA, QB, QC, and QD, respectively.

192

3)   Go through **Options → Create Netlist**, **Options → Integrity Test** and **Options → Export Netlist...** from the menu. Then, close the **Schematic Editor** window.

4)   You can perform a simulation before you download the design onto your target board. The shift-right and shift-left operations are illustrated in Figure 16.3(a)(b) below.

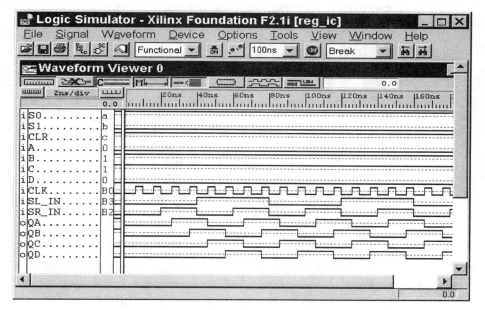

(a) When S0=1, S1=0, and CLR=1, the SR_IN signal is shifted right one clock-period on the positive edge (rising edge) of CLK.

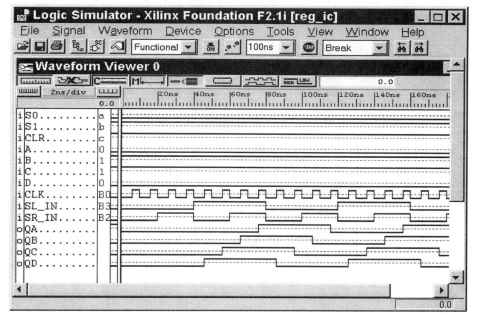

(b) When S0=0, S1=1, and CLR=1, the SR_IN signal is shifted left one clock-period on the positive edge (rising edge) of CLK.

Figure 16.3

5) Implement the design and then download the bitstream onto your target board.

6) Test the functions of the 74194 module and complete the following table.

Inputs

CLK	$\overline{\text{CLR}}$	S1	S0	SR_IN	SL_IN	A, B C, D	Function
X	0	X	X	X	X	X	
X	1	0	0	X	X	X	
↑	1	1	0	X	0	X	
↑	1	1	0	X	1	X	
↑	1	0	1	0	X	X	
↑	1	0	1	1	X	X	
↑	1	1	1	X	X	0101	
↑	1	1	1	X	X	1100	

Table 16.2 Truth Table for the 74194 Universal Shift Register

7) Determine the following for the 74194:

a) Does the input CLR depend on the CLK signal? Is the operation synchronous or asynchronous?

b) Does each shift operation take place at the ↑ or the ↓ edge of the clock pulse?

c) Why do you suppose a parallel load is called a "jam entry"?

d) If ABCD = 1011, CLR = 1, LOAD = 0, SR_IN = 0, S1 = 0, and S0 = 1,what will the outputs $Q_A Q_B Q_C Q_D$ be after two clock pulses?

e) If ABCD = 0100, CLR = 1, LOAD = 0, SL_IN = 1, S1 = 1, and S0 = 0 what will the outputs $Q_A Q_B Q_C Q_D$ be after two clock pulses?

Checked by _____ Date _____

Section II. Build a 16-bit Register Using Two 74164 Modules

You will build a 16-bit register in schematic mode. But before drawing the schematic, look at the diagram of the 74164 (Figure 16.4) and answer the three questions below:

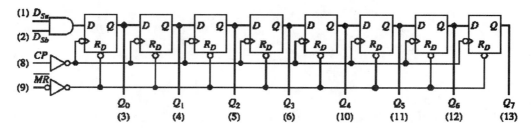

Figure 16.4 Logic diagram for 74164 module.

1)   How do you feed the serial input?

2)   Is the reset ($\overline{MR}$) synchronous or asynchronous?

3)   How should the two 74164 modules be connected?

After you have completed the answers of the above questions, build the design in Schematic mode. Verify your design using simulation tools. Your simulation result should look similar to the waveforms shown in Figure 16.5 below.

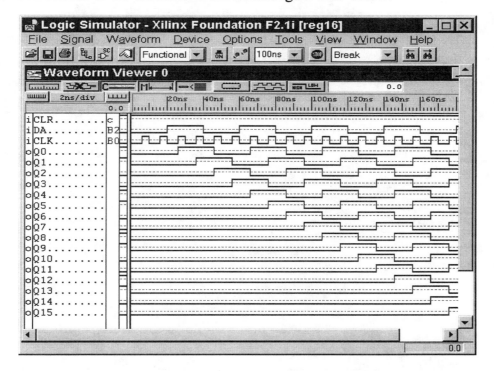

Figure 16.5 Simulation result for the 16-bit register design.

When you finish the simulation, answer the following questions:

4)   When performing a shift operation, what should you do with the RESET input?

5)   How many clock pulses are needed to shift 1 data bit from a data input (A or B) to the output of the last stage (Q_{15})?

6)   What options do you have when using output buffers in the schematic design?

Checked by _____ Date _____

Section III.   Build a Recirculating Shift Register using a 74194 Module

Recirculating shift registers are also called ring counters. In this part of the experiment, we will construct a ring counter that is capable of shifting right and left using the 74194 module introduced in Section I of this experiment.

1)   Create a new project called **ring4** in schematic mode.

2)   In the **Project Manager** window, create a schematic diagram similar to Figure 16.6 below. Note that no terminals are connected to the SLI or SRI inputs.

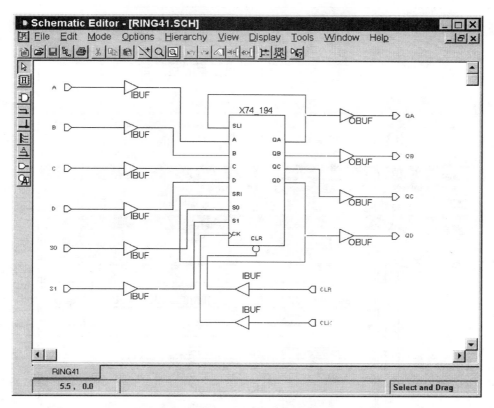

Figure 16.6 Schematic diagram for project **ring4**.

3)   Go through **Options → Create Netlist, Options → Integrity Test** and **Options → Export Netlist...** from the menu. Then, close the **Schematic Editor** window.

4)   Run a simulation to verify that the you can recirculate data using shift-right and shift-left operations. Refer to Tables 16.1 and 16.2 for the proper logic levels to assign to CLR, S0, and S1. Note that you must load the data on inputs A, B, C, and D into the register before the recirculating starts. Typical simulation results are illustrated in Figure 16.7 (a) and (b). You can use a pattern other than 1000 for A,B,C,D as long as you can see the recirculation process clearly.

197

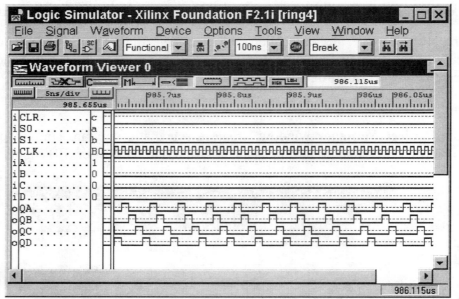

(a) Recirculating shift-right operation.

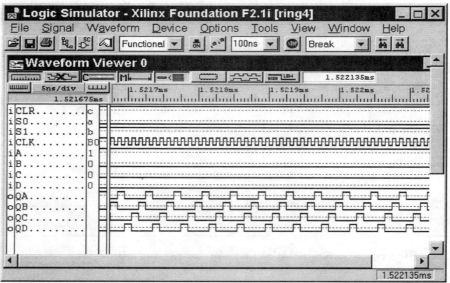

(b) Recirculating shift-left operation.

Figure 16.7

5) Assign pins to the inputs and outputs on the schematic design.

6) Go through **Options → Create Netlist**, **Options → Integrity Test** and **Options → Export Netlist...** from the menu. Then, close the **Schematic Editor** window.

7) Download the bitstream into the CPLD chip on your target board and test the operations.

Checked by _____ Date _____

Section IV.   Build A Johnson Counter Using a 74194 Module

1)   Edit the schematic diagram in Figure 16.6 to match the diagram shown in Figure 16.8 and save it in *johnson.sch* file.

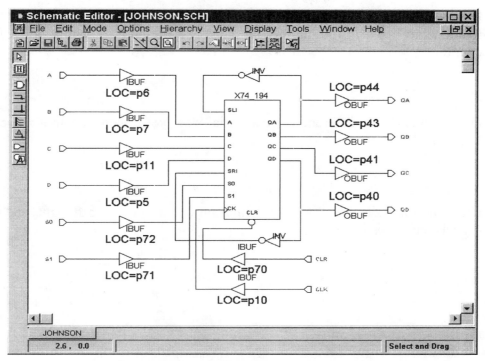

Figure 16.8 Schematic diagram for *johnson.sch*

2)   Simulate the Johnson counter design to verify its operation.   Your simulation results should look similar to Figure 16.9.

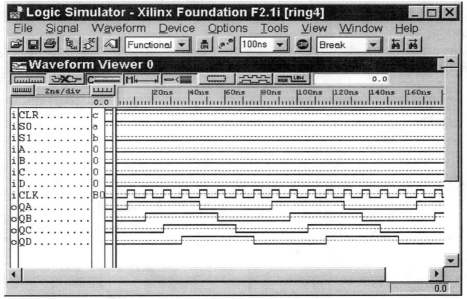

Figure 16.9 Simulation results for the Johnson counter.

3) Implement and download the design to your target board. Test the design on the target board by loading different initial inputs into ABCD and seeing the patterns on the LED outputs as the bits are shifted.

4) How many different states does this Johnson counter have? Can you find the relationship between the number of states and the number of FFs for a Johnson counter?

5) Does it make any difference what initial state you assign the Johnson counter? Explain.

6) Draw the logic diagram for a 4-bit Johnson counter using shift-right, and with an asynchronous reset and load.

7) Assume that you start the above Johnson counter at 0000. Analyze the operation of the Johnson counter and draw the timing waveforms for the outputs, along with the CLK signal, for 10 clock periods.

Checked by _____ Date _____

200

QUESTIONS:

1) Given an 8-bit ring counter, what is the maximum number of counts it can have?

2) Draw the block diagram for a 4-bit recirculating shift-right register using D FFs with asynchronous SET and CLEAR.

3) For the ring counter of question 2), how do you initialize it to $Q_A Q_B Q_C Q_D = 1011$?

4) Using the initial condition given in 3), draw timing waveforms to show the register contents together with the clock for a complete cycle of the counter.

5) Draw a block diagram that shows how to make an 8-bit recirculating shift-right register using the 74164 module. Show the connections to all the inputs.

6) Draw a block diagram to show how to make an 8-bit Johnson counter using the 74164 module and necessary logic gate(s). You need to show the connections to all the inputs.

7) Name two applications for ring-counters and Johnson-counters.

CPLD EXPERIMENT 17:
Sequential Design in HDL: Registers

OBJECTIVES:

- Use both VHDL and ABEL to design a loadable shift register.
- Compare and contrast VHDL with ABEL.
- Gain additional experience with sequential circuit design in an HDL.

MATERIALS:

- Xilinx Foundation Software, student or professional edition V1.5 or higher.
- IBM or compatible computer with Pentium processor or equivalent, 64 M-byte RAM or more, and 3 G-byte or larger hard drive.
- PLDT-1 board by RSR Electronics Inc., XS95 board and XStend Board by XESS Corp., or a similar board with an XC95108 device.

PROCEDURE:

We have examined the universal synchronous counter design in VHDL and ABEL. To further study sequential circuit design in HDL, we will demonstrate a loadable shift register design in both VHDL and ABEL in this experiment.

Section I. An 8-bit Loadable Shift Register in VHDL

1) Create a new project called *regs8* in HDL mode. Your **New Project** window should look like Figure 17.1. Click on the **OK** button to go to the **Project Manager** window.

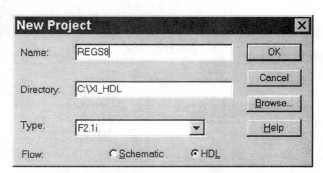

Figure 17.1 New Project window for *regs8*.

2) Click on the **HDL Editor** button on the **Design Entry** tab. In the **HDL Editor** selection window, select the **Use HDL Design Wizard** radio button and click on the **OK** button to go to the **Design Wizard** window. In the **Design Wizard** window, click on the **Next** button. Then, select the **VHDL** radio button and click on the **Next** button in the **Design Wizard – Language** window to get to the **Design Wizard – Name** window. In this window, type the name *reg8* for the *reg8.vhd* file and click on the **Next** button to go to the **Design Wizard – Ports** window.

3) In the **Design Wizard – Ports** window, define the I/O ports: CLK, SD, LOAD, CLR, DIN[0:7], and DOUT[0:7]. When this is done, click on the **Finish** button to go to the **HDL Editor** window.

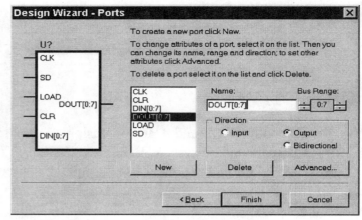

Figure 17.2 I/O ports for *reg8.vhd*.

4) The skeleton of the *reg8.vhd* file is generated. To get a template from the Language Assistants, choose **Tools → Language Assistant** from the menu in the **HDL Editor** window to get into the **Language Assistant –VHDL** window. In this window, click on the "+" next to **Synthesis Templates** and highlight the item **Shift Registers**. To view the items within **Shift Registers**, click on the "+" next to it, you will see three items. Since we want the register with parallel load ability, select the first one on the list: **Loadable Shift Register**. The VHDL codes for this loadable shift register are displayed in the right pane. Position the cursor in your **HDL Editor** window below the line <<Enter your statements here>>, click on the **Use** button to copy the template. Then, return to your **HDL Editor** window. The template is inserted into the skeleton *reg8.vhd* file as shown in Figure 17.3 (a) (b).

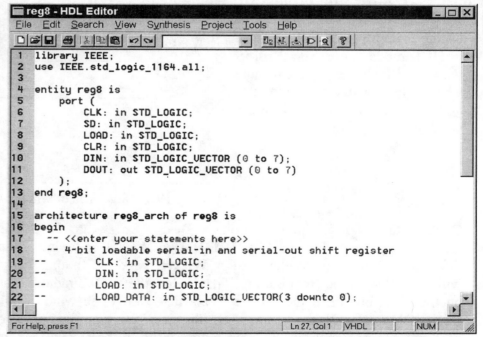

Figure 17.3 (a) VHDL template for the loadable shift register is inserted into the skeleton of the *reg8.vhd* file, line 1 to line 22.

204

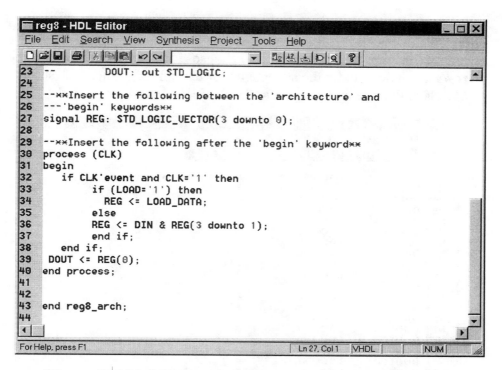

```
23 --          DOUT: out STD_LOGIC;
24
25 --**Insert the following between the 'architecture' and
26 ---'begin' keywords**
27 signal REG: STD_LOGIC_VECTOR(3 downto 0);
28
29 --**Insert the following after the 'begin' keyword**
30 process (CLK)
31 begin
32    if CLK'event and CLK='1' then
33        if (LOAD='1') then
34          REG <= LOAD_DATA;
35        else
36        REG <= DIN & REG(3 downto 1);
37        end if;
38    end if;
39  DOUT <= REG(0);
40 end process;
41
42
43 end reg8_arch;
44
```

Figure 17.3(b) VHDL template for the loadable shift register is inserted into the skeleton of the *reg8.vhd* file, line 23 to line 44.

5)   The template is designed for a 4-bit loadable shift register without a RESET operation. Therefore, we must modify it to 8-bit and add the library for using the STD_LOGIC_VECTOR type and the codes for a RESET function. Figures 17.4 (a) and (b) illustrate the *reg8.vhd* file after modification.

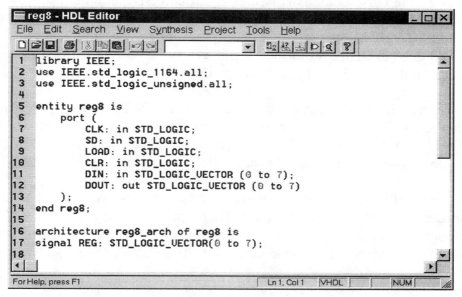

```
1 library IEEE;
2 use IEEE.std_logic_1164.all;
3 use IEEE.std_logic_unsigned.all;
4
5 entity reg8 is
6     port (
7        CLK: in STD_LOGIC;
8        SD: in STD_LOGIC;
9        LOAD: in STD_LOGIC;
10       CLR: in STD_LOGIC;
11       DIN: in STD_LOGIC_VECTOR (0 to 7);
12       DOUT: out STD_LOGIC_VECTOR (0 to 7)
13     );
14 end reg8;
15
16 architecture reg8_arch of reg8 is
17 signal REG: STD_LOGIC_VECTOR(0 to 7);
18
```

Figure 17.4 (a) 8-bit loadable shift register, line 1 to line 18.

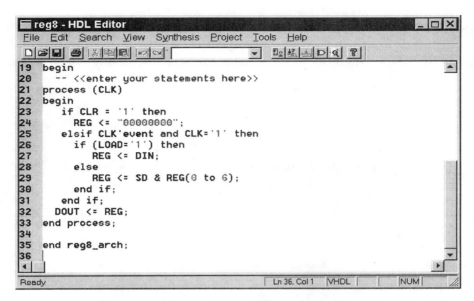

```
reg8 - HDL Editor
File  Edit  Search  View  Synthesis  Project  Tools  Help

19  begin
20    -- <<enter your statements here>>
21  process (CLK)
22  begin
23     if CLR = '1' then
24        REG <= "00000000";
25     elsif CLK'event and CLK='1' then
26        if (LOAD='1') then
27           REG <= DIN;
28        else
29           REG <= SD & REG(0 to 6);
30        end if;
31     end if;
32     DOUT <= REG;
33  end process;
34
35  end reg8_arch;
36

Ready                                    Ln 36, Col 1   VHDL          NUM
```

Figure 17.4 (b) 8-bit loadable shift register, line 19 to line 36.

Some of the VHDL code is similar to that in the universal synchronous counter in CPLD experiment 15. You may have noticed the new syntax we used on line 29 shown in Figure 17 (b). This statement represents the shift operation where the "&" symbol is used to combine the serial data bit (SD) with the REG(0 to 6) to form the new contents of the REG. In other words, the new bits SD, DIN0, DIN1, DIN2, ... , DIN6 will replace the old bits DIN0, DIN1, DIN2, ... , DIN7, respectively.

6)   To check if there is any syntax error, choose **Synthesis → Check Syntax** from the menu in the **HDL Editor** window.

7)   Choose **Project → Create Macro** from the menu. In the **Set initial target** window, make sure *XC9500, 95108PC84,* and *20* are in the editor boxes **Family, Part,** and **Speed**, respectively. Click on the **OK** button. If everything is OK, you will see the pop-up window indicating the symbol was created successfully. Click on the **OK** button.

8)   The macro *reg8* is now created. Upon returning to your **Project Manager** window, go back to the **HDL Editor** window and choose **Project → Add to Project** from the menu. Exit to the **Project Manager** window.

9)   Synthesize the macro by clicking on the **Synthesis** button. When the contents of the editor boxes in the **Synthesis/Implementation settings** window are consistent with the design, click on the **Run** button. Both the **Design Entry** and **Synthesis** buttons now show green checkmarks.

10) Use the simulation tools to verify your VHDL design.

Checked by _____ Date _____

11) To test the design on the PLDT-1 board, add the pin assignment to the ***regs8.ucf*** file. To do this, right click on **regs8** in the left pane of the **Project Manager** window. In the pull-down menu, choose **Edit Constraints**. In the **Report Browser** window, enter the statements for the pin assignments as shown in Figure 17.5. Save this file and go back to the **Project Manager** window.

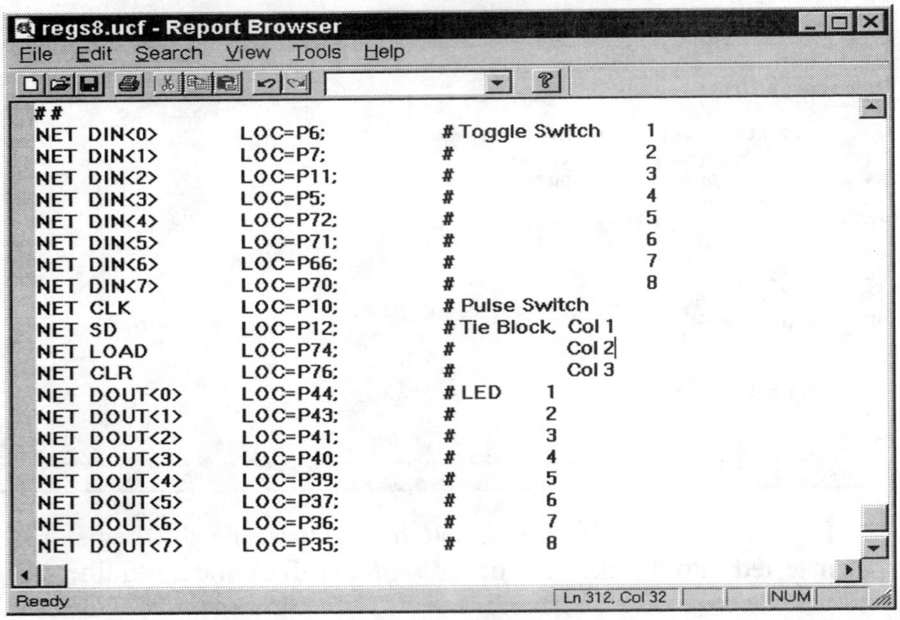

```
regs8.ucf - Report Browser                          _ □ ×
File   Edit   Search   View   Tools   Help

 □ ☞ 🖫   🖨  🗶 🗎 🗎   ↶ ↷ [              ▼]   ?|

##                                                                ▲
NET DIN<0>        LOC=P6;       # Toggle Switch      1
NET DIN<1>        LOC=P7;       #                    2
NET DIN<2>        LOC=P11;      #                    3
NET DIN<3>        LOC=P5;       #                    4
NET DIN<4>        LOC=P72;      #                    5
NET DIN<5>        LOC=P71;      #                    6
NET DIN<6>        LOC=P66;      #                    7
NET DIN<7>        LOC=P70;      #                    8
NET CLK           LOC=P10;      # Pulse Switch
NET SD            LOC=P12;      # Tie Block  Col 1
NET LOAD          LOC=P74;      #                Col 2
NET CLR           LOC=P76;      #                Col 3
NET DOUT<0>       LOC=P44;      # LED       1
NET DOUT<1>       LOC=P43;      #           2
NET DOUT<2>       LOC=P41;      #           3
NET DOUT<3>       LOC=P40;      #           4
NET DOUT<4>       LOC=P39;      #           5
NET DOUT<5>       LOC=P37;      #           6
NET DOUT<6>       LOC=P36;      #           7
NET DOUT<7>       LOC=P35;      #           8        ▼
◄ |                                              ►
Ready                          | Ln 312, Col 32 |      | NUM |
```

Figure 17.5 Pin assignment in ***regs8.ucf*** file for PLDT-1 board.

Note that VHDL requires you to assign the pins for the elements within a vector in the same order that they are declared in the VHDL program. For example, we declared the DIN vector as DIN(0 to 7) in our VHDL program, so we must make the pin assignment for DIN(0) first, followed by DIN(1), DIN(2), ..., etc. Recall that in the VHDL program of CPLD experiment 15, we had the D vector declared as D(3 downto 0) and we made the pin assignment for D(3) first followed by the assignments for D(2), D(1), and finally, D(0).

Since there are only eight toggle switches on the PLDT-1 board, we can use pins 12, 74, and 76 (on column 1, 2, and 3 of Tie Block 1) to represent inputs SD, LOAD and CLR, respectively. If your breadboard does not have additional switches, you can use jumpers to apply 0s (GND) and 1s (+5V) to the tie block.

12) Implement the design and download the bitstream onto your target board. To test the design, first clear the register. Then apply 1 to the SD input and shift it through the register to the output. Then change the SD input to 0 and shift it through the register. Shift some other bit patterns through the register.

13) Answer the following questions for your design:

a) Is RESET a synchronous or an asynchronous operation?

b) How do you modify the program to trigger the clock with a negative edge?

c) What is the purpose of line 3 in the VHDL program?

Checked by _____ Date _____

Section II. An 8-bit Loadable Shift Register in ABEL

1)   Create a new project called *regst8* in schematic mode. Your **New Project** window should look like Figure 17.6. Click on the **OK** button to go to the **Project Manager** window.

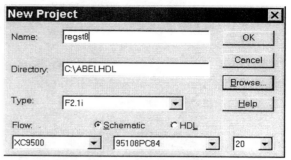

Figure 17.6 New Project window for *regst8*.

2)   In the **Project Manager** window, click on the **HDL Editor** button on the **Design Entry** tab. The **HDL Editor** selection window will appear. Select the **Use HDL Design Wizard** radio button, and then click on the **OK** button.

3)   In the **Design Wizard** window, click on the **Next** button. In the **Design Wizard – Language** window, choose the **ABEL** option and click on the **Next** button. Type the name *regt8* for the file *regt8.abl* in the editor box of the **Design Wizard – Name** window, and then click on the **Next** button.

4)   You should now have the **Design Wizard – Port** window. In this window, you will specify the inputs and outputs of the 8-bit loadable shift register. Repeat the instructions of Section I, step 3. Click on the **Finish** button to go to the **HDL Editor** window. You should see the skeleton of *regt8.abl* shown in Figure 17.7.

```
regt8.abl - HDL Editor                                    _ □ ×
File   Edit   Search   View   Synthesis   Project   Tools   Help

 1   module regt8
 2   Title 'regt8'
 3
 4   Declarations
 5
 6   CLK PIN;
 7   SD PIN;
 8   LOAD PIN;
 9   DIN0..DIN7 PIN;
10   DIN = [DIN0..DIN7];
11   CLR PIN;
12   DOUT0..DOUT7 PIN istype 'reg';
13   DOUT = [DOUT0..DOUT7];
14
15   "   <<add your declarations here>>
16
17   Equations
18
19   "   <<add your equations here>>
20
21   end regt8
22

For Help, press F1                    Ln 1, Col 1    ABEL        NUM
```

Figure 17.7 Skeleton of *regt8.abl*.

209

5)	In the **HDL Editor** window, choose **Tools → Language Assistants** from the menu. In the **Language Assistant – ABEL** window, click the "+" next to the **Synthesis Templates** and highlight **Load Register**. The ABEL code for the loadable 4-bit register with asynchronous RESET should appear in the right pane. Position the cursor below *<<Add your equations here>>* in the **HDL Editor** window and click on the **Use** button in the **Language Assistant – ABEL** window. The template is added to the *regt8.abl* in the **HDL Editor** window as illustrated in Figure 17.8 below.

```
 regt8.abl - HDL Editor                                    _ □ X
 File   Edit   Search   View   Synthesis   Project   Tools   Help

19  "   <<add your equations here>>
20  " 4-bit parallel load register with asynchronous reset
21  "          CLK           pin;
22  "          ASYNC         pin;
23  "          LOAD          pin;
24  "          DIN3..DIN0    pin;
25  "          DOUT3..DOUT0 pin istype 'reg';
26  "
27  "          DIN  = [DIN3..DIN0];
28  "          DOUT = [DOUT3..DOUT0];
29
30  equations
31
32  DOUT.CLK = CLK;
33  DOUT.ACLR = ASYNC;
34
35  when LOAD then DOUT := DIN;
36              else DOUT := DOUT.FB;
37
38  |
39  end regt8
40
 Ready                          Ln 38, Col 1   ABEL        NUM
```

Figure 17.8 The template for a 4-bit loadable shift register is added to *regt8.abl*.

6)	Since the template is for a 4-bit loadable shift register and we want 8 bits, we must make changes. Also, we need to assign I/O pins in the *regt8.abl* file. The modified program is shown in Figure 17.9. Note that we are using the same input and output variables and the same pin assignment we used in the VHDL design in Section I of this experiment. The CLR input in our design is equivalent to the ASYNC input in the template. Lines 1 to 4 are comments added to the ABEL program and are not shown in the Figure 17.9. The code is self-explanatory.

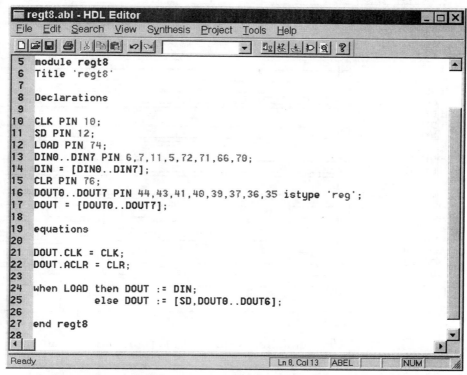

```
regt8.abl - HDL Editor                                    _ □ X
File  Edit  Search  View  Synthesis  Project  Tools  Help

 5  module regt8
 6  Title 'regt8'
 7
 8  Declarations
 9
10  CLK PIN 10;
11  SD PIN 12;
12  LOAD PIN 74;
13  DIN0..DIN7 PIN 6,7,11,5,72,71,66,70;
14  DIN = [DIN0..DIN7];
15  CLR PIN 76;
16  DOUT0..DOUT7 PIN 44,43,41,40,39,37,36,35 istype 'reg';
17  DOUT = [DOUT0..DOUT7];
18
19  equations
20
21  DOUT.CLK = CLK;
22  DOUT.ACLR = CLR;
23
24  when LOAD then DOUT := DIN;
25           else DOUT := [SD,DOUT0..DOUT6];
26
27  end regt8
28

Ready                              Ln 8, Col 13   ABEL        NUM
```

Figure 17.9 ABEL code for the register design, lines 5 to 28.

7)   To check if there is any syntax error, choose **Synthesis → Check Syntax** from the menu in the **HDL Editor** window.

8)   In the **HDL Editor** window, choose **Project → Create Macro** from the menu. Click on the **OK** button. The macro *regt8* is created.

9)   In the **HDL Editor** window, choose **Project → Add to Project** from the menu. That will include this macro in the project *regst8* so that we can perform a simulation. Save the *regt8.abl* file and exit to the **Project Manager** window.

10) Synthesize the macro by clicking on the **Synthesis** button. Make sure the editor boxes in your **Synthesis/Implementation settings** are consistent with your project. Then, click on the **Run** button. Both the Design Entry and Synthesis buttons should show green checkmarks.

11) Use the simulation tools to verify the functions of the loadable register design. Your results should be as shown in Figure 17.10 below.

Checked by _____ Date _____

211

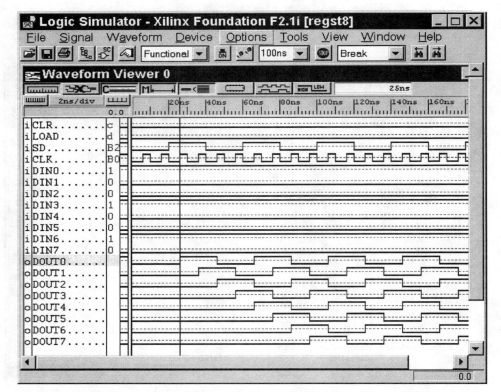

Figure 17.10 Simulation results for the right-shift operation.

12) Implement the design and download the bitstream to your target board. Test the design.

13) Determine the following:

a) The register design uses right-shift operation. How do you modify the program to shift left?

b) What changes and additions do you have to make so that the program can shift both right and left? Try to modify the design in Section II or create a new design and verify it using simulation tools.

Checked by _____ Date _____

QUESTIONS:

1)   Use the Language Assistants available in VHDL to write a complete VHDL program for an 8-bit serial-in serial-out shift register.

2)   Use the Language Assistants available in ABEL to write a complete ABEL program for an 8-bit serial-in serial-out shift register.

3)   Based on your experience in answering questions 1) and 2), compare and contrast VHDL and ABEL. If you prefer one HDL over the other, explain why.

CPLD EXPERIMENT 18:
Timing Circuits: Oscillators & One-Shots

OBJECTIVES:

- Examine the operation of common digital timing circuits.
- Design, build, and test a simple digital one-shot.

MATERIALS:

- Xilinx Foundation Software, student or professional edition V1.5 or higher.
- IBM or compatible computer with Pentium processor or equivalent, 64 M-byte RAM or more, and 3 G-byte or larger hard drive.
- PLDT-1 board by RSR Electronics Inc., XS95 (V1.2) board by XESS Corp., or a similar board with an XC95108 device.
- Oscilloscope and TTL-level square wave generator
- ICs: 555, 74HC00.
- Resistors: two 10k resistors, one 100k resistor.
- Capacitors: 0.001 µF, 0.01 µF, and 0.1 µF.

DISCUSSION:

We have looked at synchronous circuits in previous lab experiments, and they always included a clock signal. In digital equipment such as computers the clock signal is obtained from an oscillator circuit. You may have seen oscillators in a linear devices or communications course. Those oscillators are designed to produce sinusoidal signals. In digital circuits, oscillators produce what some (imprecisely) call a "square-wave". The output of a digital clock oscillator is a unipolar signal as shown in Figure 18.1 below, and is often described as being "TTL Level" when V_{CC} is 5V. Note that the duty-cycle is not necessarily 50%. Digital oscillators are sometimes called *astable multivibrators*.

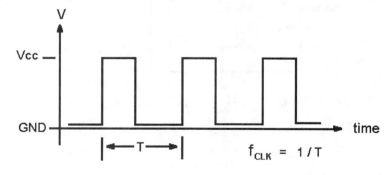

Figure 18.1 Typical output from a digital oscillator.

An oscillator has a fixed period T after which the waveshape repeats the cycle. The clock frequency is the inverse of the clock period (f = 1 / T), and is measured in Hertz (Hz). One Hz is one cycle per second. While oscillators provide a continuous time reference, a single pulse of a fixed time duration is sometimes required. Circuits that can produce single pulses on demand are called *mono-stable multivibrators*. It is common to call them *one-shots* since they produce one pulse per trigger.

Digital oscillators can be built in several ways. A common circuit is the one given in Figure 18.2 below. It uses a quartz crystal (XTAL) to get an exact frequency with very little drift. A problem with such circuits is that they don't always start; but you can buy crystal oscillator modules that do. Another way is to use an R-C controlled IC such as the venerable 555 chip as shown in Figure 18.3 below.

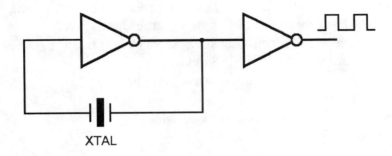

Figure 18.2 Crystal-controlled digital oscillator.

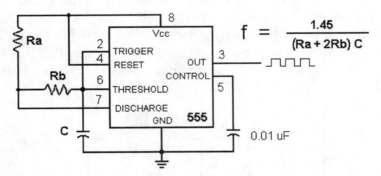

$$f = \frac{1.45}{(Ra + 2Rb)\,C}$$

Figure 18.3 555 Oscillator.

216

Most one-shot circuits use an R-C time constant to determine the width of the pulse produced. A typical circuit using CMOS gates is shown in Figure 18.4 below. It works because the R-C circuit delays the signal from the output of the first inverter to the input of the second one. Both AND gate inputs are high during that delay. There are also integrated circuits designed as one-shots; they too require an R and a C to set the pulse width.

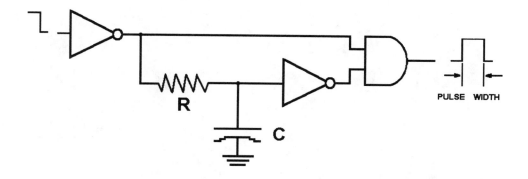

Figure 18.4 One-shot using gates & an RC time constant.

In a VLSI chip like a CPLD, there is enough available logic to allow us to build a one-shot without an RC time constant. An example is shown in Figure 18.5 below.

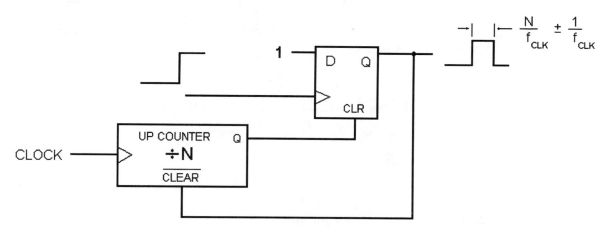

Figure 18.5 One-shot using a counter.

A common use for one-shots is to de-bounce inputs from mechanical switches.

PROCEDURE:

555 Oscillator

1) For the circuit of Figure 18.3, calculate the oscillation frequency using Ra = Rb = 10 k and C = 0.001 uF. Record your answer below.

$$f = \underline{\hspace{2cm}} Hz$$

2) Build the 555 oscillator using the same component values that you used above. Measure both the frequency and the duty cycle (DC) and record them below.

$$f = \underline{\hspace{2cm}} Hz$$

$$DC = \underline{\hspace{2cm}} \%$$

3) Calculate the percent difference between calculated frequency and measured frequency and record it below.

$$\% \, DIF = \frac{f_{CALC} - f_{MEAS}}{f_{CALC}} \times 100\%$$

$$\% \, DIF = \underline{\hspace{3cm}} \%$$

4) Is the percent difference within the tolerance of the resistors and capacitors? If not, describe what other factors could have affected the frequency.

Checked by \underline{\hspace{6cm}}   Date \underline{\hspace{3cm}}

One-Shot Using Gates & RC Delay

1) Build the circuit of Figure 18.4 on a solderless breadboard using the 74HC00 IC together with a 10k resistor and a 0.01 uF capacitor. Set your function generator to 1 kHz and use the TTL-level output to generate a train of trigger signals on the input.

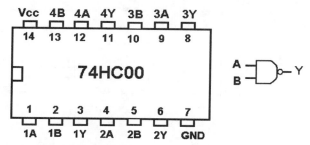

Figure 18.6 Quad-2 NAND Gate Pin-out

2) Use your oscilloscope in dual-trace mode to view the input trigger and the output pulse simultaneously. Vary the frequency of function generator while observing the pulse width. What affect does the frequency have on the pulse width? Measure the pulse width and record it here:

measured PW = _____

3) The pulse width for this kind of circuit is given by: $\mathbf{PW = K \times RC}$
where K is a constant. From the measured value of the pulse width, calculate the value for K for this circuit and record it here:

K = _____

4) Using the value for K you just obtained from the data, calculate what the pulse width should be for these values of R and C. Then use each value in your circuit and measure the pulse width. Record your results in the table below, and recalculate K for each combination of R and C. Does it change?

R OHMS	C uF	CALCULATED PW	MEASURED PW	CALCULATED VALUE FOR K
10k	0.01			
20k	0.01			
10k	0.001			
20k	0.001			
100k	0.001			
100k	0.001			

Table 18.1 Pulse Width and Calculated K Value

Checked by _____ Date _____

One-Shot using a Counter

1) We will use the XILINX software to design and implement a circuit similar to Figure 18.5 above. In the **Schematic Editor**, draw a circuit similar to Figure 18.7 below.

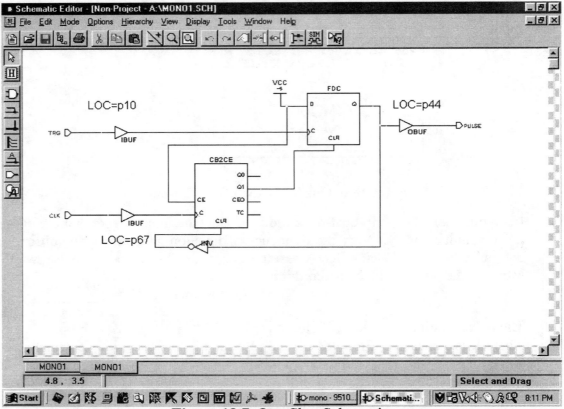

Figure 18.7 One-Shot Schematic.

2) Down-load the design to the target board. The trigger (TRG) can be a momentary push-button of a toggle switch. The pulse output (PULSE) should connect to an LED. Drive the counter's clock input (CLK) from a function generator set to 10 Hertz.

3) Activate the trigger several times and verify that the LED shows a pulse each time. Increase the clock frequency to 20 Hertz and repeat. Verify that the pulse is shorter at 20 Hz than at 10 Hz. If possible, use an oscilloscope in single-sweep mode to measure the pulse width.

4) The width of the pulse can vary by up to one clock period (why?). To reduce the variation in pulse width while producing the same width pulse, increase the number of stages in the counter and increase the clock frequency accordingly. Modify your design to include another CB2CE and repeat step 3. Is there less "jitter" in the pulse width?

Checked by _____ Date _____

220

QUESTIONS:

1) Given a 555 and a .001 uF capacitor for C, find standard resistor values for Ra and Rb to get as close as possible to a 10 kHz frequency of oscillation. Would the value C = 0.0022 make the task easier or more difficult? Could you write a computer program to do such calculations?

   OPTIONAL: Write a program as described above. Attach it to this sheet.

2) Use a book or the Internet to find out what *contact-bounce* (*switch-bounce*) is. Summarize your findings here:

3) Find at least two uses for one-shots other than debouncing, and describe them here:

4) Explain why the circuit of Figure 18.5 has a "jitter" of one clock pulse.

5)   Use the Internet or other resource to find a company that manufactures clock oscillator modules for use in digital systems. Summarize the specifications of a typical unit and show how the clock oscillator could be used in a digital circuit.

CPLD EXPERIMENT 19:
Digital to Analog Converter

OBJECTIVES:

- Examine the operation of digital to analog conversion (DAC) circuitry.
- Design, build, and test a simple DAC.

MATERIALS:

- Xilinx Foundation Software, student or professional edition V1.5 or higher.
- IBM or compatible computer with Pentium processor or equivalent, 64 M-byte RAM or more, and 3 G-byte or larger hard drive.
- PLDT-1 board by RSR Electronics Inc., XS95 (V1.2) board by XESS Corp., or a similar board with a 95108 device.
- Oscilloscope with X10 probe.
- TTL-level square wave generator.
- 100k/200k R-2R DIP resistor (Bourns part # 4116R-R2R-104 or equivalent) or six 200 kΩ and four 100 kΩ resistors, 1/4 W, 1% tolerance.
- 100k resistor, 1/4 W, 5%.
- The following capacitors: 0.001 μF, 0.01 μF, 0.1 μF.

DISCUSSION:

We want to have digital equipment that can interface to us users. But the information we users can understand is in the form of analog signals such as sound and sight. As we know, what we hear is a fluctuation in air pressure and what we see is a fluctuation in light. We can use transducers such as speakers and CRTs to convert analog voltages into things we can hear and see. But if we want our digital systems to talk to us and show us pictures, then we need a way to convert discrete bits into continuous analog signals. We need a digital to analog converter, or DAC.

We are going to use a resistor network called an "R-2R Ladder" as shown in Figure 19.1 below. Each switch represents one bit, with S0 being the LSB and S4 being the MSB. A switch set to V_{REF} is a binary 1 while a switch set to ground is a 0. The analog output is at the point labeled V_X . When all the switches are at ground V_X is zero. If only S4 is switched from ground to V_{REF} we can use Ohm's Law to calculate that $V_X = V_{REF} \div 2$.

In the same manner, we can calculate that S3 switched to V_{REF} by itself will produce $V_X = V_{REF} \div 4$, S2 by itself will produce $V_X = V_{REF} \div 8$, S1 by itself will produce $V_X = V_{REF} \div 16$, and S0 by itself will produce $V_X = V_{REF} \div 32$. An R-2R ladder is a linear circuit, so from superposition we know that the effect of multiple switches is the sum of the effects of each individual switch. So if all the switches were put to V_{REF}, we would get: $V_X = (1/2 + 1/4 + 1/8 + 1/16 + 1/32) V_{REF} = 31/32 V_{REF} = 0.96875 V_{REF}$. The voltage produced by S0 is the value, or weight, of one bit. So 1 bit = 0.03125 V_{REF}.

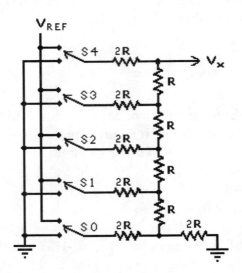

Figure 19.1 R-2R Ladder Network

In this experiment we will use a 4-stage up-counter to produce the digital data. We will connect the outputs of the flip-flops to an R-2R network, and clock the counter as we observe the analog output on an oscilloscope.

PROCEDURE:

Section 1: Counter Design

Refer back to CPLD Experiment 13 on asynchronous counters. Use the Foundation Software to design a 4-bit up-counter. Assign the clock input to pin 10. For the PLDT-1, assign the outputs as follows: Q3 (MSB) to pin 12, Q2 to pin 74, Q1 to pin 76, and Q0 (LSB) to pin 77; all on tie-block 1. Use the simulation tools to verify its operation.

Checked by _____   Date _____

Section 2: R-2R Ladder

Build the R-2R network on a solderless breadboard. If you have a DIP resistor pack (Figure 19.2), insert it into the breadboard and use four bits as shown. If you are using discrete resistors, then mount them onto the breadboard as shown in Figure 19.3 below.

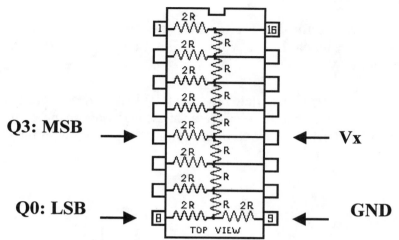

Figure 19.2 DIP resistor pack ladder.

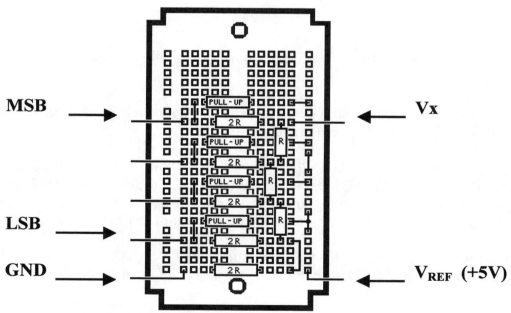

Figure 19.3 Discrete resistor 4-bit ladder with pull-ups.

Checked by _____   Date _____

225

Since we are using the voltages from the CPLD directly without using a precision reference voltage, the steps may not be of equal size. To compensate, use 10 kΩ pull-up resistors on the Q outputs of the flip-flops (see Figure 19.3).

Section 3: Analog Output

Connect the Q outputs of the flip-flops to the R-2R network using wires from the target board to the solderless breadboard. Use your oscilloscope with a X10 probe to observe the voltage on the output (V_X).

Set your TTL level square-wave oscillator to about 10 kHz and connect it to the clock input on the target board. You should observe a "staircase" waveshape similar to Figure 19.4 below with $2^4 = 16$ steps. Is it *monotonic* (do the steps go up with the count)? Is it *linear* (are all the steps the same size)?

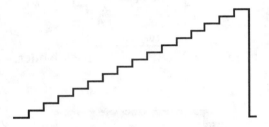

Figure 19.4 Staircase output at V_X: a saw-tooth waveform.

Vary the clock frequency and observe the result. Note the relationship between the clock frequency and the frequency of the analog signal.

Modify the connections to the R-2R network so that Q0 is disconnected and Q1 becomes the LSB. Outputs Q2 and Q3 are all moved down one position on the R-2R ladder. See Figure 19.5 below. You should now observe a waveshape as in Figure 19.4 except that the "step size" should be twice as large and the number of steps should be $2^3 = 8$.

On the next page, measure V_x for each count of the 3-bit DAC and fill in Table 19.1. Then draw a graph of the DAC output (V_x) versus binary input for the 3-bit counter.

Checked by _____ Date _____

Section 4: Linearity

DIGITAL INPUT		OUTPUT VOLTAGE
BINARY	DECIMAL	
000	0	
001	1	
010	2	
011	3	
100	4	
101	5	
110	6	
111	7	

Table 19.1 DAC Input / Output

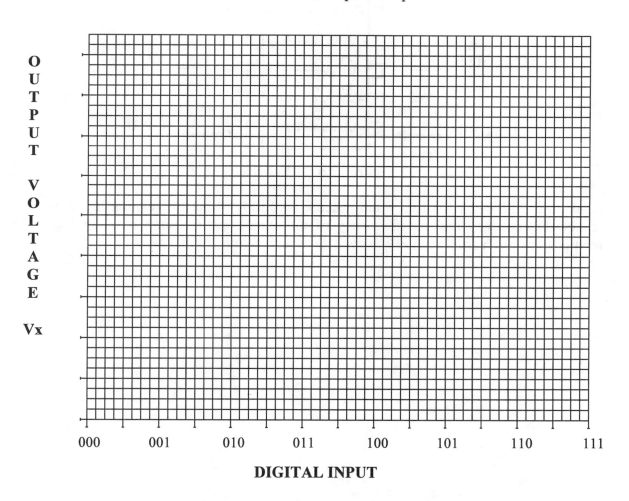

Figure 19.5 Plot of data (to be filled in by student)

After you plot your data, draw a straight line from the point (000, Vmin) to (111, Vmax). Does your data fall on the line? At what values of digital input are the voltages farthest from the straight line? Comment on how *linear* this DAC seems to be. What do you think affects linearity?

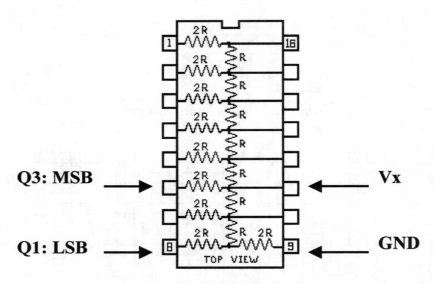

Figure 19.6 3-Bit ladder

Again, vary the clock frequency and observe the result. Note the relationship between the clock frequency and the frequency of the analog signal. In what ways do the relationships differ from section 3 above? Summarize your results here:

Checked by _____   Date _____

Section 5: Low-Pass Filtering

The output of a saw-tooth generator typically would not have the steps of the staircase output shown in Figure 19.4 above. It would look more like this:

Figure 19.7 Saw-tooth waveform.

The steps on the output of a DAC represent frequency components that are much higher than the frequency of the saw-tooth signal. Those high frequencies can be thought of as noise. To remove that noise, we will use a low-pass filter (LPF) between the output voltage (V_X) and the oscilloscope. A simple LPF can be made from an R-C circuit:

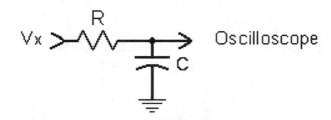

Figure 19.8 R-C low pass filter (LPF).

Insert the filter using R = 100k and C = 0.001 µF and observe the output wave-form. Then replace C with the 0.01 µF capacitor and see what effect it has. Then replace C with the 0.1 µF capacitor and sketch two cycles of the waveshape on the next page.

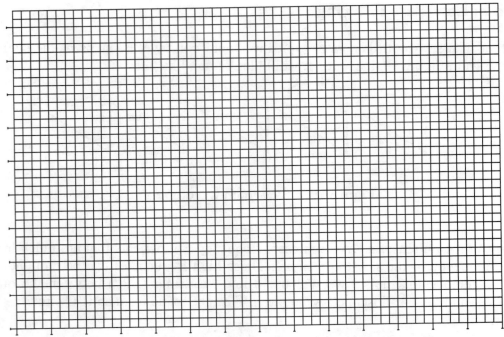

Figure 19.9 Two cycles of waveshape to be drawn by student.

Modify the counter design to make it a down-counter. Again, observe the waveshape. How has it changed? Sketch one cycle of the waveform here:

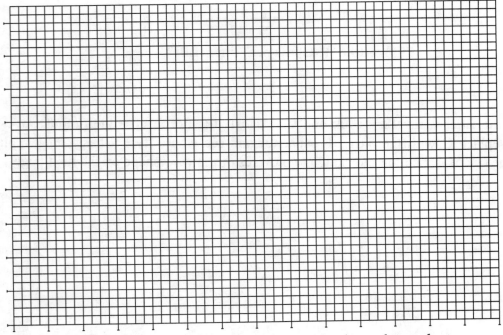

Figure 19.10 One cycle of waveshape to be drawn by student.

Checked by _____ Date _____

QUESTIONS:

1) Write an equation that gives the analog frequency as a function of the clock frequency and the number of bits being used.

2) How could you modify the counter design so that the analog waveshape would be a triangle wave? Draw the modified counter design.

3) With five bits, you would have $2^5 = 32$ steps. If the reference voltage was 10 Volts, what would be the "size" of one step? (i.e., find the Volts / step.) Show the calculation.

4) What percentage of the maximum output voltage is one step? Explain how the resistor tolerance relates to that percentage.

USE FOR NOTES OR AS A WORK-SHEET

NAME _____ **DATE** _____

CPLD EXPERIMENT 20:
Analog to Digital Converter

OBJECTIVES:

- Examine the operation of analog to digital conversion (ADC) circuitry.
- Design, build, and test a simple ADC.
- Plan a procedure from an outline of the experiment.

MATERIALS:

- Xilinx Foundation Software, student or professional edition V1.5 or higher.
- IBM or compatible computer with Pentium processor or equivalent, 64 M-byte RAM or more, and 3 G-byte or larger hard drive.
- PLDT-1 board by RSR Electronics Inc., XS95 (V1.2) board by XESS Corp., or a similar board with an XC95108 device.
- Digital Voltmeter.
- Solderless breadboard, 10k linear potentiometer, 1k resistor, R2R ladder network from CPLD Experiment 19.
- Integrated circuits: LM311 comparator, 74xx00 Quad-2 NAND gate (7400, 74LS00, 74HC00, 74HCT00).

DISCUSSION:

Computers and other digital systems are used to store, manipulate, and process information in the form of 1s and 0s that we call bits. But most of the real-world information we want to process is in the form of smoothly varying physical signals in a natural medium. A typical example of a physical signal is voice. It can be converted from air pressure fluctuations (its natural medium) into an AC voltage by a transducer such as a microphone. But the AC voltage is the continuously changing analog of the voice signal; it's not 1s and 0s. To convert a signal such as voice into bits, the analog voltage must go through the process of digitization using an analog-to-digital converter (ADC or A/D).

The key idea of analog to digital conversion is *sampling*. Sampling is a process which, in effect, takes "snapshots" of the analog voltage at discrete points in time. Each voltage sample is then converted into a binary number that is proportional to its amplitude. The number of samples taken per second is the *sampling rate*. The number of bits used to represent each sample is the *resolution* of the ADC. In any given application, we need to determine appropriate values for both the resolution and the sampling rate. Let's look at the resolution first.

Resolution

Suppose we have an analog voltage that ranges between 0 and 4 Volts. And suppose we have an ADC with a resolution of 2 bits. With 2 bits we have $2^2 = 4$ possible binary numbers which can represent four voltage levels as shown in Table 20.1 below. Notice that we can't represent a value of 4 Volts!

ANALOG VOLTAGE	MSB	LSB
0 Volt	0	0
1 Volt	0	1
2 Volts	1	0
3 Volts	1	1

Table 20.1 2-Bit Digital Resolution of Analog Voltage

But what if a voltage sample is, say, 1.2 Volts? The answer is that the ADC will convert it to the binary number 01, the same as if it was 1.0 Volts. So the information contained in the truncated 0.2 Volts is lost. In fact, any sample that is greater than 0.5 Volts and less than 1.5 Volts will be converted to binary 01. Look at table 20.2 below:

NOMINAL INPUT	VOLTAGE RANGE	MSB	LSB
0.0 V	$0.5 > V \geq 0.0$	0	0
1.0 V	$1.5 > V > 0.5$	0	1
2.0 V	$2.5 > V > 1.5$	1	0
3.0 V	$4.0 \geq V > 2.5$	1	1

Table 20.2 Voltage Ranges for 2-Bit Resolution

And what if the sample were 1.5 Volts? Then the output of the ADC could be either 01 or 10; flip a coin. What we are looking at here is quantization error, a sort of noise produced by the analog to digital conversion process. We can calculate the quantization error as a percentage using the voltages: (1 V/step) ÷ (4 V max) $\Rightarrow$ 25%. But we don't need to know the voltages, we just need to know the bits of resolution: $1 \div 2^2 \Rightarrow 25\%$

If we change to a 3-bit ADC, we will get Table 20.3 below. Again, for inputs on the edge of two ranges, such as 1.25 Volts, the binary output could go either way.

NOMINAL INPUT	VOLTAGE RANGE	MSB	BIT	LSB
0.00 V	$0.25 > V \geq 0.00$	0	0	0
0.50 V	$0.75 > V > 0.25$	0	0	1
1.00 V	$1.25 > V > 0.75$	0	1	0
1.50 V	$1.75 > V > 1.25$	0	1	1
2.00 V	$2.25 > V > 1.75$	1	0	0
2.50 V	$2.75 > V > 2.25$	1	0	1
3.00 V	$3.25 > V > 2.75$	1	1	0
3.50 V	$4.0 \geq V > 3.25$	1	1	1

Table 20.3 Voltage Ranges for 3-Bit Resolution

Three bits has given us a "finer grain", meaning better resolution. We still have quantization noise, but now it is $1 \div 2^3 \Rightarrow 12.5\%$ (or $0.5\,V \div 4\,V \Rightarrow 12.5\%$). The more bits in the ADC, the less the percent error as shown in Table 20.4 below:

ADC RESOLUTION	PERCENT ERROR
4 Bits	$1 / 2^4 \Rightarrow 6.25\ \%$
8 Bits	$1 / 2^8 \Rightarrow 0.39\ \%$
12 Bits	$1 / 2^{12} \Rightarrow 0.024\ \%$
16 Bits	$1 / 2^{16} \Rightarrow 0.0015\ \%$

Table 20. 4 Error percentage & Resolution

The 0.0015% error of a 16-bit ADC translates into approximately 15 μV per Volt.

Sampling Rate

The question of how many samples per second are required has an answer that is simple to state (but a little harder to prove): the sample rate must be faster than twice the highest frequency in the analog signal. This is the *Sampling Theorem*:

$$SR > 2 \times f_H$$

where SR is the sample rate in samples per second, and f_H is the highest frequency component in the analog signal that is to be digitized. The value of $2f_H$ is called the *Nyquist rate*.

To understand what is meant by "the highest frequency component", we need to know that any AC signal (such as voice or video) that is *not* a pure sine-wave can be expressed mathematically as a sum of sine-waves called *component frequencies*. That group of sine-waves is called the signal's *frequency spectrum* (or just *spectrum*). (You can find the details of this in many textbooks under the heading "Fourier Analysis".)

For example, the components of a square-wave with a period (T) of 1 millisecond would contain a *fundamental frequency* sine-wave of $f_0 = 1 / T = 1$ kHz, but would also contain in its spectrum sine-waves of 3 kHz, 5 kHz, 7 kHz, and higher odd frequencies. As the frequencies of the component go up, the amplitudes of the components go down quickly, so above some frequency we can ignore them. See Figure 20.1 below.

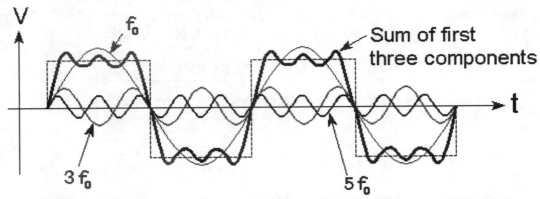

First Three Frequency Components of a Square-Wave

Figure 20.1 Sine-wave components of a square-wave.

Since the sample rate is usually fixed for an ADC, before we digitize a complex AC signal, we need to pass it through a low-pass filter (a *pre-filter*) so that it becomes a *band-limited* signal. Otherwise, the signal may contain frequency components that are higher than half the Nyquist rate. If you sample a signal too slowly, you will create distortion (called "fold-over distortion" or "aliasing") and lose information.

A/D Conversion Techniques

Many ADCs use a *comparator*. A comparator is an analog integrated circuit with two inputs and one output. Figure 20.2 shows an LM311 device. Its role is to indicate which of two analog input voltages is the greater. If the input marked with the plus sign (+) is greater, the output voltage will be HIGH at the logic-1 level (typically +5 Volts). Otherwise, the output is LOW at the logic-0 level (typically ground). A typical value for V_{CC} is 12 Volts.

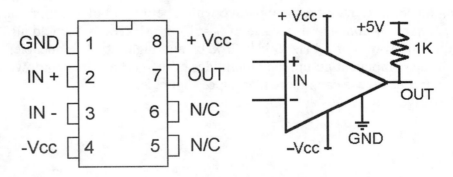

Figure 20.2 LM311 Comparator.

To make an ADC, we will apply the analog voltage to one input of a comparator and the output from a DAC to the other input. We need a digital circuit to generate a sequence of binary numbers until the comparator indicates we have found the right one.

There are two ways to generate the sequence of binary numbers. The simpler, but slower, way is to use an up-counter. For each sample, the counter is cleared and then allowed to count up. The output of the DAC will be a staircase as we saw previously in CPLD Experiment 19. We apply the analog sample to the comparator's (+) input and the output of the DAC to its (-) input.

The output of the comparator is used to control the counter's clock as shown in Figure 20.3 below. When the staircase from the DAC is one bit's worth higher than the sample, the comparator "flips" and stops the count. The number in the counter corresponds to the value of the analog sample.

The second, and faster, way to generate the binary number is a method called *successive approximation*. In successive approximation, the output of a register (the *SAR* register) drives the DAC. The register starts out cleared. Then the register MSB is set HIGH. If the comparator flips, the MSB is set back to LOW; otherwise, it is left HIGH. The procedure is then repeated with the next highest bit of the register, and then the next, until the LSB is reached. At that point, the register holds the binary number for that sample.

In this experiment, we will use the up-counter method.

PROCEDURE:

Section 1: Build the 4-Bit Up-Counter

Refer back to CPLD Experiment 19. Design, build, and test a 4-bit DAC. Use a toggle switch to control the counter reset. Use two separate pins for each output: one for the DAC and one for an LED to display the count.

Section 2: Build the A/D Control Circuit

On a solderless breadboard, build the circuit shown in Figure 20.3 below using an LM311 IC comparator and one gate from a 7400 quad-2 NAND gate. The Vcc for the comparator can be in the range of 5 to 15 Volts. Use a 10 Hz TTL-level signal from a function generator as the clock source. Use an LED to display the state of the comparator's output. Use a toggle switch to control the counter's CLEAR input.

237

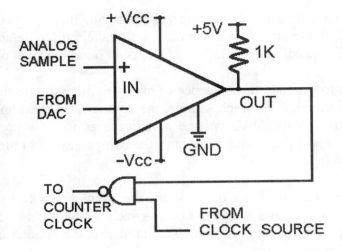

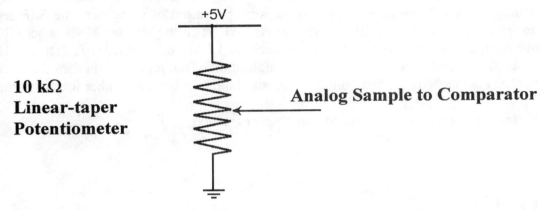

**10 kΩ
Linear-taper
Potentiometer**

Analog Sample to Comparator

Figure 20.3 ADC control circuit & analog signal source.

As shown in Figure 20.3 above, use a 10k potentiometer to generate DC voltages as the analog samples. Connect the control circuit to the target board. Activate the counter's CLEAR and verify that the LED on the output of the comparator is off. Deactivate the CLEAR. When the comparator LED lights, the conversion is done.

Checked by _____ Date _____

Section 3: Test the ADC with Sample Voltages

On the next page we will be drawing a pair of graphs. Here we will collect the data for those two graphs.

Conversion

Use a digital voltmeter to measure both the analog sample and the output of the DAC. Read the digital value on the LEDs. For each specified sample voltage, when the conversion is done, record the values in Table 20.5 below.

| SAMPLE VOLTAGE | | DAC | | DIGITIZED VALUE | |
NOMINAL	MEASURED	VOLTAGE		BINARY	DECIMAL
0.00 V	V	V			
0.50 V	V	V			
1.00 V	V	V			
1.50 V	V	V			
2.00 V	V	V			
2.50 V	V	V			
3.00 V	V	V			
3.50 V	V	V			
4.00 V	V	V			
4.50 V	V	V			
5.00 V	V	V			

Table 20.5 Digitization Data for Specified Voltage Inputs

Linearity

Linearity is a measure of how well the ADC output tracks the analog input. For a discussion of linearity, refer to your textbook, and/or use a search engine on the Internet. Make measurements as you did for Table 20.5 except adjust the analog input voltage until you get the specified binary output. Fill in Table 20.6 with the measurements.

| DIGITAL OUTPUT | | MEASURED |
BINARY	DECIMAL	SAMPLE VOLTAGE
0001	1	
0010	2	
0100	4	
0110	6	
1000	8	
1010	10	
1100	12	
1110	14	

Table 20.6 Digitization Data for Specified Binary Outputs

Checked by _____ Date _____

Section 4: Graphs

1)   From the data in Table 20.5, draw a graph showing the measured output voltage from the DAC on the Y-axis vs. the nominal values for the analog sample voltage on the X-axis.

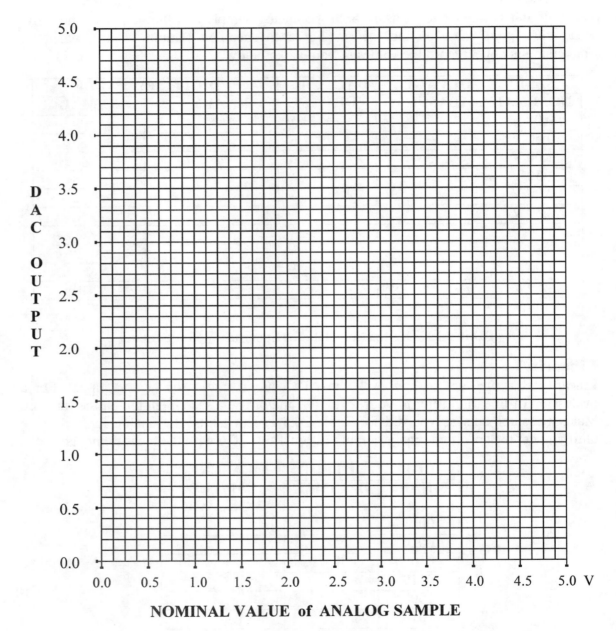

Figure 20.4 Graph of data from Table 20.5.

2) From the data in table 20.6, draw a graph showing the digitized output as a decimal value on the horizontal axis and the corresponding measured value of the sample voltage.

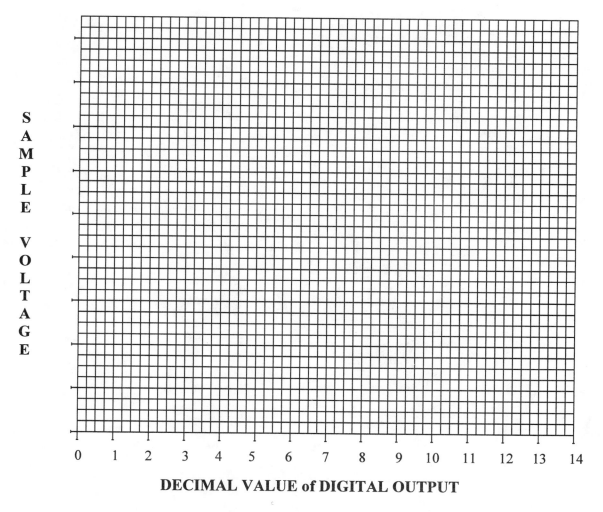

DECIMAL VALUE of DIGITAL OUTPUT

Figure 20.5 Graph of data from Table 20.6.

Section 5: Analysis

In the following space, explain what the two graphs you just drew tell you about:

1) the tracking of the digital output with the analog input
2) the linearity of the ADC built in this experiment

Checked by _____ Date _____

QUESTIONS:

1)   What is meant by the term "resolution" ?

2)   How does resolution differ from accuracy ?
        Hint: $1/2 = 0.5$ is low resolution and $1/2 = 0.235752$ is high resolution. Accuracy?

3)   What would be the Nyquist rate for sampling an audio signal with a bandwidth of 200 Hz to 4000 Hz ?

4)   From your text, or other source, explain what "aliasing" is. Use a sketch as part of your explanation.

5)   Explain the concept of a signal's spectrum. Use a sketch as part of your explanation.

6)   Based on your data, what voltage would correspond to the digital number 1010?

7)   From question 1), what would be the range of voltages corresponding to 1010?

8)   Based on your data, what digital number would correspond to 2.80 Volts?

9)   From question 3), how much would the voltage have to drop to produce a digital number one bit less than the answer to 3)?

10)   Sketch a design for an SAR type of ADC.

APPENDICES

APPENDIX A:
REFERENCES

---

1. XILINX On-Line Technical support: http://support.xilinx.com/support/support.htm

2. *THE PRACTICAL XILINX® DESIGNER LAB BOOK*
   by David Van den Bout,   Prentice-Hall (1999) ISBN 0-13-021617-8

3. *DIGITAL DESIGN: PRINCIPLES & PRACTICES 3rd Ed.*
   by John F. Wakerly,   Prentice-Hall (2000) ISBN 0-13-769191-2

4. *MODELING, SYNTHESIS and RAPID PROTOTYPING with the VERILOG HDL*
   by Michael D. Ciletti,   Prentice-Hall (1999) ISBN 0-13-977398-3

5. *DIGITAL SYSTEMS: PRINCIPLES and APPLICATIONS 7th Ed.*
   by Ronald J. Tocci, Prentice-Hall (1998) ISBN 0-13-700510-5

6. *DIGITAL ELECTRONICS: A PRACTICAL APPROACH, 5th Ed.*
   by William Kleitz , Prentice-Hall (1999) ISBN 0-13-769274-9

7. *DIGITAL FUNDAMENTALS 7th Ed.*
   by Thomas Floyd, Prentice-Hall (2000)   ISBN 0-13-080850-4

8. *LOGIC and COMPUTER DESIGN FUNDAMENTALS 2nd Ed.*
   by M. Morris Mano & Charles R. Kime , Prentice-Hall (2000) ISBN 0-13-012468-0

9. *PARALLEL PORT COMPLETE*
   by Jan Axelson, Lakeview Research (1996) ISBN 0-96-508191-5

APPENDIX B:
GLOSSARY

- A -

ABEL
A high-level design language (HDL) and compilation system produced by Data I/O Corporation.

actions
In state-machines, actions are HDL statements that are used to make assignments to output ports or internal signals. Actions can be executed at several points in a state diagram. The most commonly used actions are state actions and transition actions. State actions are executed when the machine is in the associated state. Transition actions are executed when the machine goes through the associated transition.

Aldec
An Electronics Design Automation (EDA) vendor. Aldec provides the Foundation Project Manager, Schematic Editor, Logic Simulator, HDL Editor, and State Editor.

aliases
Aliases, or signal groups, are useful for probing specific groups of nodes.

analyze
The Foundation Express process in which design source-files are examined for correct syntax.

architecture
The common logic structure of a family of programmable integrated circuits. The same architecture can be realized in different manufacturing processes. An example of a Xilinx architecture is the XC9500 family of CPLDs.

- B -

back-annotation
The translation of a routed or fitted design to a timing simulation netlist.

bitstream
The binary file with the .jed extension used by the JTAG Programmer to configure the CPLD.

black box instantiation
An instantiation where the synthesizer is not given the architecture or modules. In Foundation, black boxes are translated with the implementation tools.

block

1. A group of one or more logic functions.
2. A schematic or symbol sheet. There are four types of these blocks:
   a) A *Composite* block indicates that the design is hierarchical.
   b) A *Module* block is a symbol with no underlying schematic.
   c) A *Pin* block represents a schematic pin.
   d) An *Annotate* block is a symbol without electrical connectivity that is used only for documentation and graphics.

Boolean

A parameter that has only two states: HIGH (True, 1) or LOW (False, 0). A mathematical equation or expression describing the inputs to outputs relationships in a digital circuit. Named after George Boole.

breakpoint

A condition for which a simulator must stop to perform simulation commands.

buffer

1. A storage element.
2. An element used to increase the drive-power of a weak signal in order to increase its fan-out.

bus

A group of nets carrying common information. In LogiBLOX, bus sizes are declared so that they can be expanded accordingly during design implementation.

- C -

CLB

Configurable Logic Block. They provide the physical support for an implemented and downloaded design in FPGAs.

component

An instantiation or symbol reference from a library of logic elements that can be placed on a schematic.

condition

A Boolean expression. If there is more than one transition leaving a state machine, you must associate a condition with each transition.

constraints

Specifications for the implementation process. There are several categories of constraints: routing, timing, area, mapping, and placement constraints.

constraints editor

A GUI tool used to enter design constraints. In Foundation 2.1, there are two constraint editors. The Express editor is available only in the Foundation Express product configuration. The Xilinx Constraints Editor is integrated with the Design Implementation tools and is available in all product configurations.

constraint file

Specifies constraints information (e.g. location and path-delay) in a textual form. An alternate method is to place constraints information on a schematic.

CPLD

A Complex Programmable Logic Device. It is an erasable logic device (integrated circuit) that can be programmed with a schematic or a behavioral design. CPLDs constitute a type of complex PLD based on EPROM or EEPROM technology. They are characterized by an architecture offering high speed, predictable timing, and simple software.

The basic CPLD cell is called a macrocell, which is the CPLD implementation of a CLB. It is composed of AND gates and is surrounded by the interconnect area. CPLDs are primarily used to support behavioral designs and to implement complex counters, complex state machines, arithmetic operations, and so forth.

CPLD fitter

The part of the software that implements designs for the XC9500 devices.

- D -

design entry tools

The set of software tools, accessible from the Project Manager, that allow a logic circuit to be described. These tools include the Schematic Editor, State Editor, and HDL Editor. Foundation Express, an embedded portion of the Foundation software package, contains the VHDL and Verilog design languages.

daisy chain

A sequence of bitstream files concatenated in one file. It can be used to program several devices that are connected in series on a circuit board.

design implementation tools

Those parts of the Xilinx software, such as the JTAG programmer, required to put a logic circuit into a device.

DRC

See physical design rule check below under P.

- E -

effort level
Refers to the optimization level the Xilinx Design System (XDS) uses to place a design into a device. The effort level settings are:

1. High, which provides the best placement but requires the longest execution time. Use high effort on designs that otherwise do not route or do not meet required performance.

2. Medium, which is the default level. It provides a trade-off between execution time and placement quality for most designs.

3. Low, which provides the fastest execution time and adequate placement results for prototyping simple (easy to route) designs. Low effort is useful if you are exploring a "design space" (several different designs) to get an estimate of final performance.

elaborate
The HDL process that combines the individual parts of a design into a single design, and then synthesizes the design.

- F -

fanout
The maximum number of specified unit loads that an output can drive.

fitter
The software that maps a PLD logic design into the target CPLD.

floorplanning
The process of choosing the best groupings and connectivity of logic in a design to optimize performance.

FPGA
Field Programmable Gate Array. A class of integrated circuits in which the logic function is defined by the designer using software after the device has been manufactured. An FPGA requires a memory chip to hold the configuration defined by the designer.

functional simulation
The process of identifying logic errors in a design before it is implemented in a device. Since timing information for the design is not available, the simulator tests the logic using unit delays. Functional simulation is usually done at the early stages of the design process.

- G -

gate
A low-level part of a digital integrated circuit. A gate is composed of a few transistors and is capable of performing primitive logic functions such as AND, OR, XOR, or NOT (inversion). Also called a switching circuit or logic circuit.

- H -

HDL
Hardware Description Language. An HDL describes circuits in a textual code, and uses a high level of abstraction to describe designs in a technology-independent manner. The two most widely used HDLs are VHDL and Verilog.

HDL Editor
Foundation's editor for ABEL, Verilog, and VHDL. It provides a syntax checker, language templates, and access to the synthesis tools.

hierarchical design
A design composed of multiple sheets at different levels of the schematic.

Hierarchy Browser
The left-hand portion of the Foundation Project Manager that displays the current design project. It also displays two tabs: Files and Versions.

- I -

implementation
The mapping, placement, and routing of a design. A phase in the design process during which the design is placed and routed. For CPLDs, the design is fitted.

instantiation
Incorporating a macro or module into a top-level design. The instantiated module can be a LogiBLOX module, a CORE-generated module, a VHDL module, a Verilog module, a schematic module, a state-machine, or a netlist.

- J -

JEDEC
The JEDEC Solid State Technology Association, formerly known as the Joint Electronic Device Engineering Council. The semiconductor engineering standardization body of the Electronic Industries Alliance (EIA).

JEDEC File
Industry standard format for the pattern information necessary to program an in-system - programmable device such as a CPLD.

JTAG
Joint Test Action Group. Specifications for the interface to program an in-system-programmable device such as a CPLD.

- L -

Language Assistant
In the HDL editor, the Language Assistant provides templates to aide in common VHDL and Verilog constructs, common logic functions, and architecture-specific features.

Library Manager
Used to perform a variety of tasks on the design entry tools libraries and their contents. The libraries contain the primitives and macros that are used to build a design.

locking
Lock placement applies a constraint to all placed components in a design. The locking option specifies that placed components cannot be unplaced, moved, or deleted.

LogiBLOX
A Xilinx design tool for creating high-level modules such as counters, shift-registers, and multiplexers.

logic
Integrated circuits that contain a logic circuit designed to manipulate or control digital signals. Logic ICs are distinct from microprocessor and memory devices.

Logic Simulator
A real-time interactive tool which can be used to carry out functional and timing simulations of a design before it is downloaded into a device. The Logic Simulator creates the equivalent of a breadboard of the design directly from the design's netlist. It can be accessed by clicking on either the Functional Simulation icon on the Simulation-phase button or the Timing Simulation icon on the Verification-phase button in the Project Manager.

- M -

mapping
The process of assigning a design's logic elements to the specific physical elements that actually implement logic functions in a device such as a CPLD.

- N -

net

1. As an abstract concept, a net is a logical connection between two or more symbol instance

   pins. After routing, the net is transformed to a physical connection called a wire.

2. An electrical connection from a single component, or between components or nets. Used in that sense, it is the same as a wire or a signal.

netlist

A text description of a circuit's connectivity. Basically, it is a list of connectors and instances. For each instance, it contains a list the signals connected to the instance terminals. Also, the netlist contains attribute information.

NGDBuild

The program that performs all the steps necessary to read a netlist file in XNF or EDIF format and create an NGD file describing the logical design. The GUI equivalent is called Translate.

NGD file

Native Generic Database file. It is an output from the NGDBuild run. It contains a logical description of the design expressed both in terms of the hierarchy used when the design was first created, and in terms of lower-level Xilinx primitives to which the hierarchy resolves.

- O -

one-hot encoding

When, in a state-machine, an individual state register is dedicated to one state. Only one flip-flop is active (hot) at any one time.

optimization

The process that decreases the device area or increases the speed of a design. Foundation allows control of a design's optimization on a module-by-module basis. Thus some modules can be optimized for area, some for speed, and some for a balance of both.

- P -

path delay

The time it takes for a signal to travel from one end of a path to the other.

PCF file

Physical Constraints File.

PDF file
Project Description File. It contains library and other project-specific information. These files have a .pdf extension, but they are *not* Adobe Acrobat files.

physical Design Rule Check (DRC)
A series of tests to discover logical and physical errors in the design.

pin
Can refer to a symbol pin or a package pin.
1. A symbol pin (or instance pin) is the connection point of an instance to a net.
2. A package pin is a physical metal contact on an integrated circuit package that carries signals into and out of the device.

- R -

radix
The base of the number system in which waveforms are displayed in a waveform viewer. Common bases are binary (base-2), octal (base-8), decimal (base-10), and hexadecimal (base-16).

- S -

Schematic Flow
A project design mode. A Schematic Flow may only have schematic designs as top-level designs, in which case the Synthesis button will not be displayed in the Project Manager. If the design contains HDL modules, the Synthesis tab displays in the upper portion of the Project Flow area.

states
The values stored in the memory elements, such as flip-flops, of a device. A specific set of logical values corresponds to each state.

state diagram
A diagram showing the states of a design, their interconnections, the inputs required to cause a state change, and the outputs of the states. States are represented by circles. Arrows going into a circle are inputs to that state, while arrows coming out of the circle are outputs from that state. Lines interconnecting the circles are the state transitions, and next to each such line is a description of what causes that transition to occur.

state machine
A set of combinatorial and sequential logic elements interconnected to operate in a predefined way to a sequence of specified inputs. The action a state machine takes depends on its current state as well as the current inputs.

state table
A tabulation showing the outputs for all combinations of current states and inputs. From any given state, the table defines the next state of a state machine for each set of inputs.

static timing analysis
An evaluation of the point-to-point delays in a design network.

static timing analyzer
A software tool that carries out the static timing analysis of a design based on the paths in that design.

status bar
An area at the bottom of a tool window. Displayed in that area is information about the commands that can be selected or the commands that are being processed.

stimulus information
Information defined at the schematic level that represents a list of nodes and vectors to be simulated in functional and timing simulations.

synthesis
The HDL design process in which each design module is elaborated, and the design hierarchy is created and linked to form a unique design implementation. Synthesis starts from a high level of logic abstraction (typically Verilog or VHDL) and automatically creates a lower level of logic abstraction using a library of primitives.

- T -

translate
The process of merging all the input netlists before the design is fitted.

- U -

UCF file
User Constraint File. Contains user-specified logical constraints.

- V -

vector
A set of input values to be applied to a design to see if the correct output is obtained. Also called a test-vector.

verification
The process of reading back the configuration from a device and comparing it to the original design to ensure that all of the design was correctly received by the device.

Verilog
An industry-standard HDL (IEEE std 1364) originally developed by Cadence Design Systems and now maintained by OVI. Verilog can be used to model a digital system at levels of abstraction ranging from the algorithmic level to the gate level. Verilog files have a .v extension.

VHDL
VHSIC HDL, where VHSIC stands for Very High Speed Integrated Circuit. An HDL that is capable of describing the concurrent and sequential behavior of a digital system with or without timing. VHDL is an industry-standard HDL (IEEE std 1076) which can be used to model a digital system at levels of abstraction ranging from the algorithmic level to the gate level. VHDL files have a .vhd extension.

- W -

wire
1. a net
2. a signal

A CAR ALARM USING SCHEMATIC-BASED DESIGN

OBJECTIVES:

- Learn how to create a new project and save a schematic design file using Xilinx Foundation Software.
- Learn how to implement a project.
- Learn how to download both JTAG and .SVF files to a target board.

MATERIALS:

- Xilinx Foundation Software, student or professional edition V1.5 or higher.
- IBM or compatible computer with Pentium processor or equivalent, 64 M-byte RAM or more, and 3 G-byte or larger hard drive.
- PLDT-1 board by RSR Electronics Inc., XS95 board and XStend Board by XESS Corp., or a similar board with an XC95108 device.

IMPORTANT: If you are using V1.5 of the Foundation software, then you need to copy the file XC95108_V2.bsd from the floppy disk to the FNDTN/DATA folder. Also, unused pins in 9500 devices are left floating unless you check the "create programmable ground pins" option in the fitter options. Selecting that option ties unused CPLD pins to ground internally.

DISCUSSION:

It is assumed that the reader is familiar with digital fundamentals. If not, then refer to any digital textbook on the subject. See references in Appendix A.

In this tutorial, you will learn how to use the Xilinx schematic design entry tools to create a simple project: a car alarm. You will start with the problem description, the truth table and the logic diagram. Then you will be guided through the process of project design, implementation, and device programming using Xilinx Foundation Software. Pay careful attention to the design procedure because it is not limited to this problem.

For convenience and clarity, we will use the following text conventions:

- Boldface and italic are used to indicate the words that you need to enter into the editor boxes on the windows (example: *INPUT*).
- Boldface is used to represent selections and mouse clicks on buttons and windows. All mouse clicks are single left-clicks unless specified otherwise.
- The PLDT-1, XS95, or whatever target CPLD board you are using will be referred to as "the target board" or as just "the board".

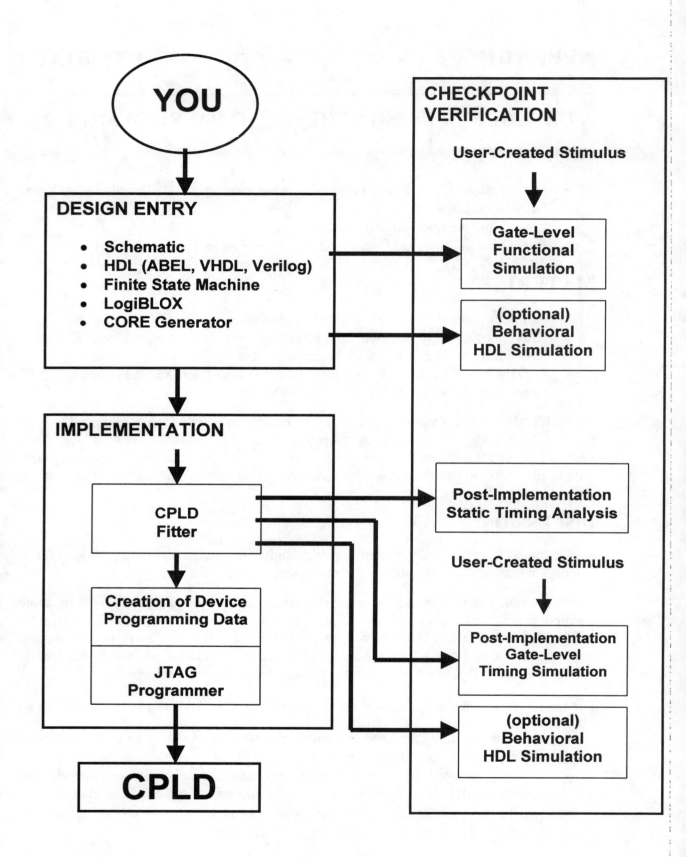

Foundation Overall Design Flow for CPLDs

Project Description

Our car alarm has three inputs: *Door*, *Ignition* and *HeadLight*. The output is the *Alarm*. The *Alarm* should sound off whenever any of the following conditions is met:

- The *Door* is OPEN while the *Ignition* is ON.
- The *Headlight* is ON while the *Ignition* is OFF.

Let us represent the status of the input/output signals with the following logic levels:

Logic Level	Door	Ignition	HeadLight	Alarm
0	CLOSED	OFF	OFF	OFF
1	OPEN	ON	ON	ON

Table T1.1 Logic Level Definition of Inputs and Output for the Car Alarm

Based on the above problem statement and the defined input/output logic levels, the truth table can be made. Notice that the truth table will require 8 rows to express all the input combinations.

Door	Ignition	HeadLight		Alarm
0	0	0		0
0	0	1		1
0	1	0		0
0	1	1		0
1	0	0		0
1	0	1		1
1	1	0		1
1	1	1		1

Table T1.2 Truth Table for the Car Alarm Problem

The simplified Boolean equation can be derived using a K-map or Boolean algebra and yields the following result:

$$Alarm = \overline{Ignition} \bullet HeadLight + Door \bullet Ignition$$

Based on the above equation, we can use one inverter, two 2-input AND gates, and one 2-input OR gate to implement this design. You will see next how to transform this Boolean equation into a working PLD circuit on the target board using Xilinx Foundation software.

PROCEDURE:

1) Create a directory called *XiTutor* in your C: drive (You may use any other name for the directory). You will save your project files for the tutorials in that directory.

2) Start the Foundation Series software by using your mouse to choose: **Start → Programs → Xilinx Foundation Series 2.1i → Project Manager** in Windows 95/98/NT. You can also start by double clicking the **Project Manager** Icon on the Windows desktop screen.

3) You should see a window similar to the **Getting Started** window illustrated on the right. *(You will not get exactly the same window since other design files may or may not be available on your hard drive).* Choose **Create a New Project** and click on the **OK** button.

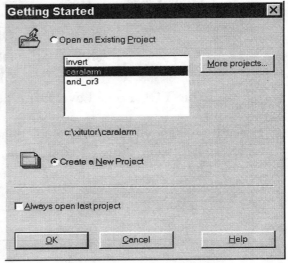

Figure T1.1 Getting Started window.

4) In the **Name** edit box, type *CarAlarm*. This will be your project file name. In the **Directory** editor box, either type or browse to get *XiTutor*. Choose **Type** as **F2.1i**. Click on **Schematic**. Make sure the **Family**, **Part**, and **Speed** are selected to be: *XC9500*, *95108PC84* and *20*, respectively. Then, click on the **OK** button.

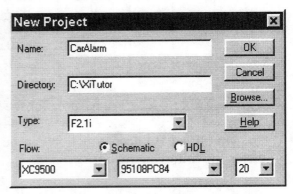

Figure T1.2 New Project window.

260

5) The new project is now created and you should see the window (Figure T1.3) shown below. As you can see, there are three main panes tiled in the **Project Manager** window. The left pane shows the source files and libraries for your design project in the **Files** tab. The **Versions** tab shows design implementation files generated by the Xilinx Foundation Software for all the versions of your design. The right pane has four tabs. The one we will use in this tutorial is the **Flow** tab.

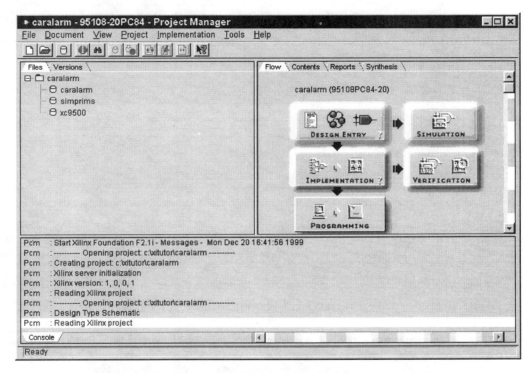

Figure T1.3 Project Manager window.

The **Flow** tab is the default choice and is shown in the window when starting and retrieving a project. It allows you to perform the steps of design, simulation, implementation, verification, and programming in the order indicated by the arrows. The bottom pane provides status information about your project such as the files created, the warnings, errors, and so forth.

6) Since we want to create a schematic-based design in this tutorial, click on the **Schematic Editor** button  on the **Design Entry** tab in the right pane. A blank Schematic Editor window (Figure T1.5) appears.

Figure T1.4 The AND button (right) on the Design Entry tab invokes the Schematic Editor.

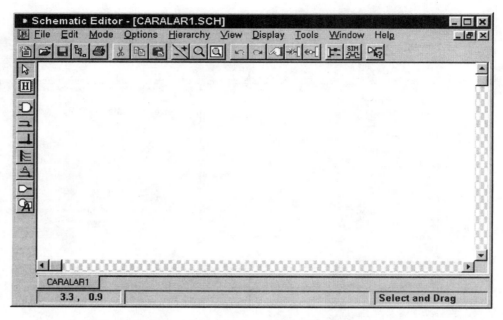

Figure T1.5 Schematic Editor window.

7) It is time now for us to draw the schematic on this window.

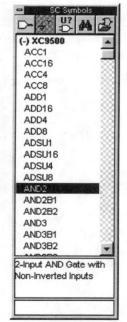

Click on the logic symbol icon 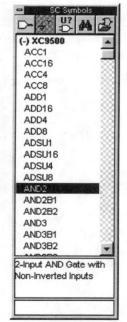 to get the **SC Symbols** window. There you can browse the logic functions and click on the symbol you need, then click on the desired location in the **Schematic Editor** window to drop it. Move the glide bar to AND2, the 2-input AND gate as explained in the message pane at the bottom of the **SC Symbols** window. Select the gate and drop it on your schematic. A 2-input AND gate symbol should now appear on the **Schematic Editor** window. Since you need two such AND gates, make a copy by clicking and dragging the AND symbol you just obtained. Select and copy the 2-input OR gate (OR2) and the inverter (INV) in the same manner. Your schematic now should look like Figure T1.7 below.

Figure T1.6 SC Symbols window.

262

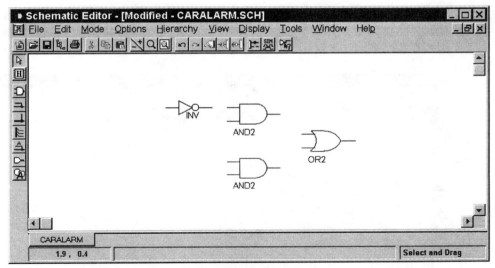

Figure T1.7 Schematic Editor window for step 7.

8) Save your schematic before something bad happens to it. Choose **File → Save As**. A window similar to Figure T1.8 should now appear. Edit the **File name** box to *caralarm.SCH* and click on the **OK** button to save the file to the CARALARM subdirectory.

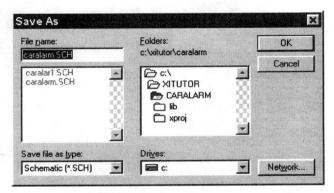

Figure T1.8 Save As window for step 8.

9) In the same manner you placed the gates, add input buffers (IBUF) and output buffers (OBUF) to your circuit. Buffers ensure that the signals attached to them will actually enter and leave the CPLD chip via the I/O pins.

10) Next, you add I/O terminals. Do this by clicking the **Hierarchy Connector** icon. Once you click on this icon, a window similar to Figure T1.9 should appear. For the first signal, type *DOOR* in the **Terminal Name** editor box. In the **Terminal Type** editor box, choose *INPUT*.

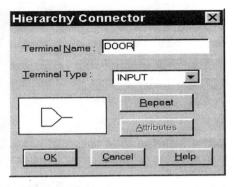

Figure T1.9 Hierarchy Connector window.

263

Repeat the process for the other two input signals: *IGNITION,* and *HEADLIGHT*. When getting the terminal for *ALARM*, choose **Terminal Type** as *OUTPUT*. The **Schematic Editor** window now should look like Figure T1.10.

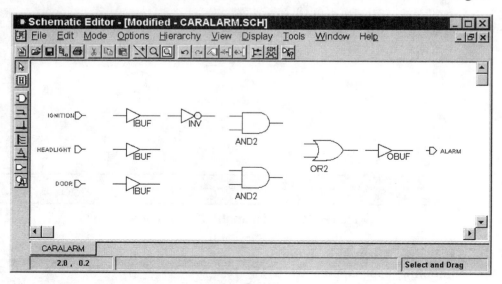

Figure T1.10 Schematic Editor window after step 10.

11) The symbols must be connected with wires. Do this by clicking on the **Draw wires** icon ![wire icon]. If a wire is drawn incorrectly, you can delete it by selecting the wire (the wire changes to red) and pressing the **Delete** key. If a wire does not end properly, you can right click the mouse and choose **Cancel** or **End net** and redraw it. When you finish all the connections, your **Schematic Editor** window should look like Figure T1.11 below.

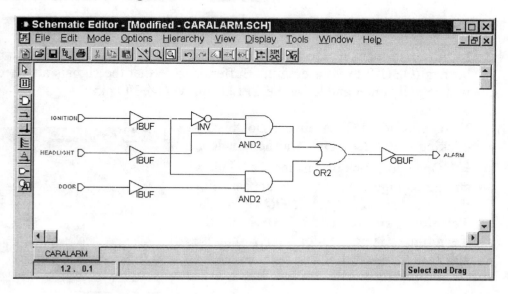

Figure T1.11 Schematic Editor window after step 11.

264

12) The Xilinx 95108PC84 CPLD chip has many pins for I/O and as well as a few for dedicated purposes. The target board is designed to use specific I/O pins for getting input signals from switches and displaying the output signals on LEDs. The pins shown in the following tables are for the PLDT-1.

Toggle Switch	1	2	3	4	5	6	7	8
CPLD Pin	6	7	11	5	72	71	66	70

Table T1.3 CPLD Pins Corresponding to the Toggle Switches on Target Board

LEDs	1	2	3	4	5	6	7	8
CPLD Pin	44	43	41	40	39	37	36	35

Table T1.4 CPLD Pins Corresponding to the LEDs on Target Board

Since you have three inputs, you will need three CPLD pins that connect to three switches. Make pin 6 = *Door*, pin 7 = *Ignition*, and pin 11 = *HeadLight*. Also, use LED 1 to display the output by making pin 44 = *Alarm*. Once you have made the pin assignment, you have to inform the CPLD chip about them. This is done as follows: Double click the input buffer for the **Door** signal. The **Symbol Properties** window should appear. Type *LOC* in the **Name** box and **P6** (pin 6) in the **Description** box (either in lower or upper case). Click on the **Add** button. This will put the line ***LOC=p6** on the message box. (You have locked the signal to that pin.) Click on the **Move** button to close this window and drag the LOC=p6 to the desired location on the schematic.

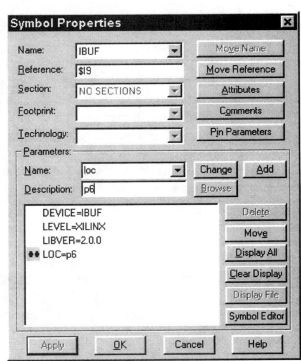

Figure T1.12 Symbol Properties window.

Do the same for all the input and output buffers. When you finish, your schematic should look like Figure T1.13 below. Note that you can adjust the symbol locations to fit the pin labels. If a wire does not appear the way you would like, you can select and drag it to change its appearance. If there are some unwanted traces left on the schematic, you can get rid of them simply by drawing a box around that portion of the schematic and then deselecting it. The unwanted traces on the schematic will disappear after you save the file and reopen it.

265

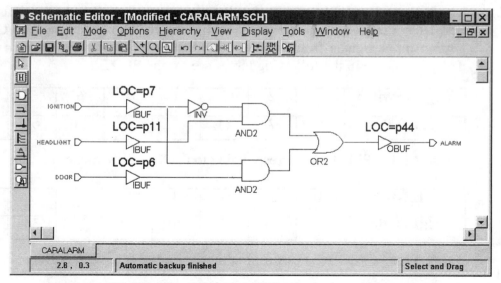

Figure T1.13 The Schematic Editor window after step 12.

13) You have completed the logic circuit design. Save your schematic by choosing **File → Save**.

14) Choose **Option → Create Netlist**. The following message window will appear. Click on the **OK** button. (You can ignore any warnings here since they will not affect your design.) This step generates an ABL binary file that describes the connections among schematic components.

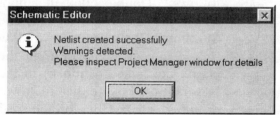

Figure T1.14 Message after Create Netlist.

15) Choose **Option → Integrity Test**. Click on the **OK** button in the message window to go back to the **Schematic Editor** window. This option performs a comprehensive analysis of the project netlists. It detects errors such as duplicated hidden net names and hanging wires terminated with bubbles at both ends.

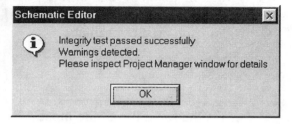

Figure T1.15 Message after Integrity Test.

266

16)   Select **Option** again, then **Export Netlist**. In the Export Netlist window, choose *Edif 200 [*.EDN]* for the **Files of type** since this is the standard netlist format for the Foundation tools. Click on the **Open** button to start the exporting process. After a few seconds, you will go back to the **Schematic Editor** window.

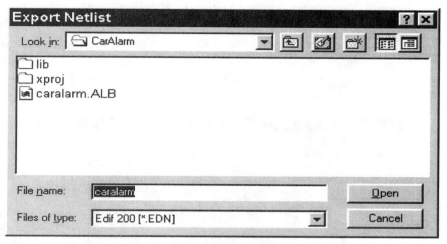

Figure T1.16   Export Netlist window.

17)   You have now completed your car alarm design in Schematic mode. Return to the **Project Manager** window by selecting **File → Exit**. There should be a green check mark on the lower right corner of the **Design Entry** tab indicating that the design has been completed successfully.

18)   The next step is to implement the schematic design. The complete implementation process consists of four stages: Translate, Fit, Timing, and Bitstream. In the **Project Manager** window, click on the **Implementation** button (or **Implementation → Implement Design** on the menu bar). Make sure the **Device, Speed, Version name** and **Revision name** editor boxes have *95108PC84, 20, ver1,* and *rev1*, respectively. Keep in mind, you must be consistent with the **Device** and **Speed** you chose at the beginning of your schematic design. Click on the **Run** button.

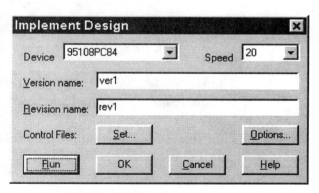

Figure T1.17   Implement Design window.

19) Now you will see the **Flow Engine** window. Expect the implementation process to take up to 15 seconds depending on the speed of your PC. When a stage is in processing, you will see "Running" in the message box underneath the stage symbol. Whenever a stage is finished without error, you will read "Completed" underneath the stage symbol and see the arrow pointed to the next stage becomes solid. If an error occurs at any stage, the process simply stops at that stage and an error message appears in the bottom message window. Select **Flow** → **Close** (or click on the Close button [■]) to go back to the **Project Manager** window.

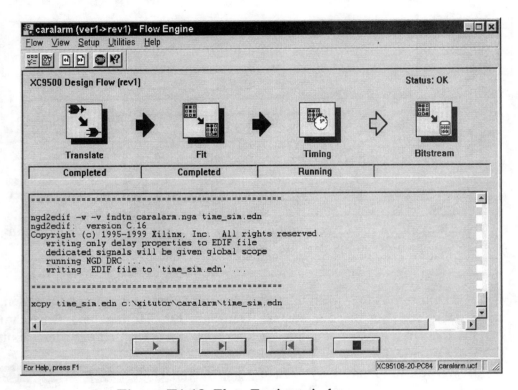

Figure T1.18 Flow Engine window.

20) The **Implementation** button should be marked with a green check, indicating that the bitstream is generated and ready to be downloaded into the 95108 chip.

21) Plug the AC adaptor into an outlet and connect it to your target board. Connect your PC to your target board with the parallel cable.

268

22) You are now ready to program the CPLD chip. Click on the **Device Programming** button (or **Tools → Device Programming** on menu bar). The **JTAG Programmer** window shown in Figure T1.19 below should appear.

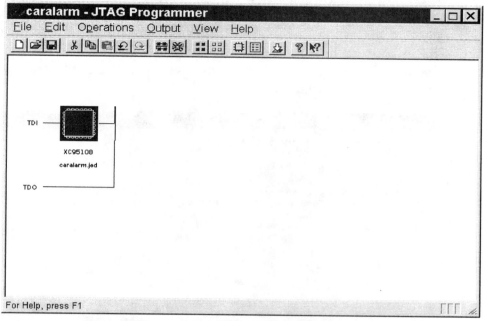

Figure T1.19 JTAG Programmer window.

23) In the **JTAG Programmer** window, choose **Operations → Program** from the menu bar, the **Options** window should pop up. In this window, select **Erase Before Programming** and **Verify**, then click on the **OK** button.

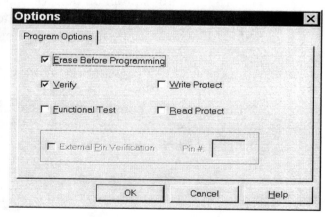

Figure T1.20 Options window.

24) *If you are using the XS95 board, skip this step and go through steps 25 to 27. If you are using the PLDT-1 board, go to step 28 directly after you finish this step.*

The download process takes a few seconds depending on the speed of your PC and the device speed you selected. During the process, you should see the **Operation Status** window. When the process is finished without any errors, the **Operation Status** window should look similar to Figure T1.21. Click on the **OK** button to go back to the **JTAG Programmer** window. In this window, choose **File → Exit**. When the message window prompts, choose **Yes**. In the **Save As** window, click on the **Yes** button to save the JEDEC file and go back to the **Project Manager** window.

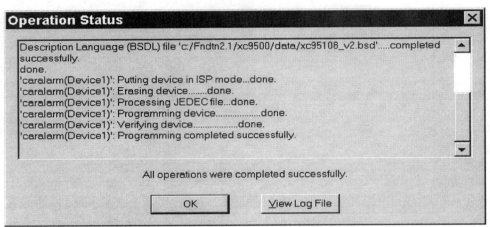

Figure T1.21   Operation Status window after step 24.

25) The XS95 board is programmed with a serial vector format file. Such files have a .svf extension. The file is downloaded to the board with a utility program called XSLOAD which can be found on the Foundation Software disk.

From the **JTAG Programmer** window, select the **Output → Create SVF file...** menu item. When the **SVF Options** window pops up (Figure T1.22), select the radio button labeled **Through Test-Logic Reset**. Then click on the **OK** button.

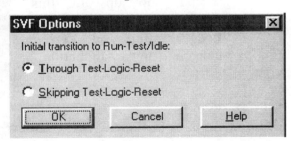

Figure T1.22 SVF Options window.

26) Next a window labeled **Create a New SVF File** pops up (Figure T1.23). In this window you can change the default names for the **Save in** folder and for the **File Name**. If there is no need to change the defaults, just click on the **Save** button. *If you changed folders, then remember where you saved the .svf file; you will need it in a minute.*

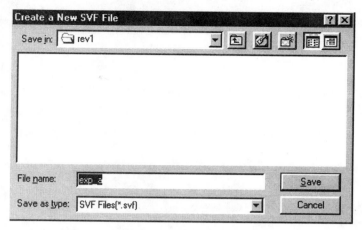

Figure T1.23 Create a New SVF File from the JTAG Programmer window.

27) Next, from the JTAG Programmer select **Operations → Program**. An **Options** window will pop up. Click on the **OK** button. The **Operation Status** window should say **SVF vector generation ... completed successfully**. You are now ready to program the target board.

Open a DOS window. If necessary, change directory to where you saved the XSLOAD program. Then, at the DOS prompt type:

XSLOAD <path>XXX.svf

Where <path> is the DOS path to where you saved the *.svf* file and **XXX.svf** is the name of the file. For example:

XSLOAD A:\EE101\LAB1.SVF

Obviously, it would make life easier if XSLOAD and your *.svf* files were in the same directory.

28) The car alarm program now "resides" in the CPLD chip. Test the program on the target board by verifying the truth table shown in the discussion part of this tutorial. Recall that we used toggle switches 1, 2, and 3 to represent the inputs *Door*, *Ignition*, and *HeadLight*, respectively while LED 1 is used to display the *Alarm* output.

271

USING SIMULATION TOOLS TO TEST A DESIGN

OBJECTIVE:

- Learn how to use simulation tools in Xilinx Foundation Software to test and verify your circuit design.

MATERIALS:

- Xilinx Foundation Software, student or professional edition, V1.5 or higher.
- IBM or compatible computer with Pentium or equivalent processor, 64 M-byte or more of RAM, and a 3 G-byte or larger hard drive.

DISCUSSION:

It is assumed that the reader is familiar with digital fundamentals. If not, then refer to any digital textbook on the subject.

In this tutorial, you will learn how to use simulation tools to verify your logic designs. Computer simulations are widely used in engineering since they can greatly shorten product development time and thereby reduce product cost. They are also invaluable tools in all areas of scientific research. The Xilinx Foundation Logic Simulator is a real-time interactive design analysis tool. It can be used with the Foundation Schematic Editor or as a stand-alone tool allowing you to interact with your design as if it were an actual hardware implementation. You can test and observe the operation of your design before you download it into the CPLD chip. We will demonstrate how to perform the design simulation using the car alarm problem that we discussed in Tutorial I.

Recall that the alarm problem is expressed by the following Boolean equation:

$$Alarm = \overline{Ignition} \cdot HeadLight + Door \cdot Ignition$$

The circuit has three inputs (*Door, Ignition, HeadLight*) and one output (*Alarm*). Refer to Tutorial I for the truth table.

PROCEDURE:

It is assumed that you have gone through Tutorial I and the caralarm files already exist in your C drive.

1)   Start the Foundation Series software by using your mouse to choose:
Start → Programs → Xilinx Foundation Series 2.1i → Project Manager in Windows 95/98/NT. You can also start by double clicking the **Project Manager** Icon on the Windows desktop screen.

2)   You should see a window that is similar to the **Getting Started** window illustrated on the right. Since you would like to simulate the *caralarm* design problem, highlight the *caralam* file and choose **Open an Existing Project** and click on the **OK** button.

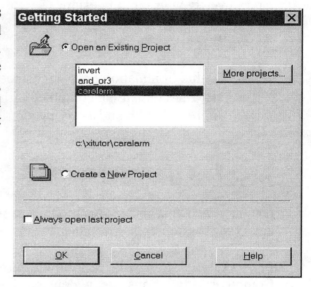

Figure T2.1 Getting Started window.

3)   You should see the **Project Manager** window next. Click on the **Simulation** button. If a message window appears, click on the **Yes** button to update the netlist from the Schematic Editor.

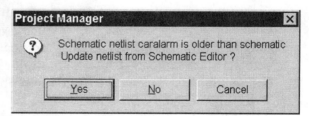

Figure T2.2 Message window.

4)   The **Logic Simulator** window shown as Figure T2.3 should appear. Observe that within the **Logic Simulator** window there is a **Waveform Viewer** window. The input and output timing waveforms will be displayed in this window.

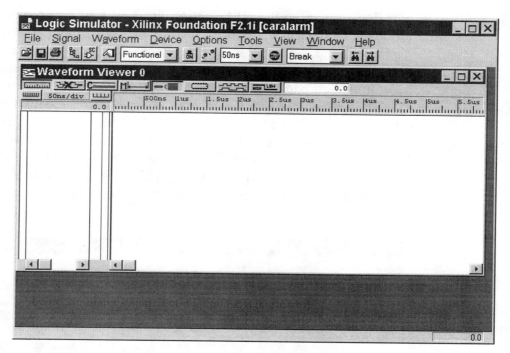

Figure T2.3 Logic Simulator window after step 4.

5) Choose **Signal** → **Add Signals** from the menu bar of the **Logic Simulator** window. The **Component Selection for Waveform Viewer** window appears on top of the **Logic Simulator** window. Observe that all inputs and output signals are listed in the Signals Selection pane. However, they are not selected yet. You may move the **Component Selection for Waveform Viewer** window to the right so that the next step can be viewed better.

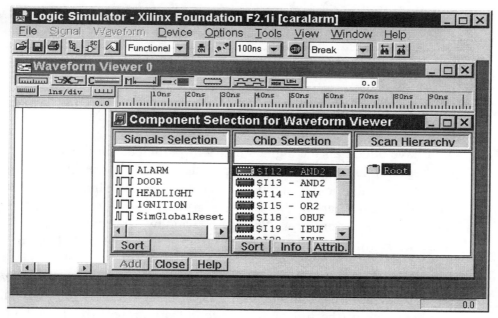

Figure T2.4 Logic Simulator window after step 5.

6) To select a signal from the **Signal Selection** list, first highlight the signal's name on the left pane of the **Component Selection for Waveform Viewer** window, and then click on the **Add** button at the bottom. When a signal is added, you should see a red checkmark on the signal selected and also the occurrence of the signal name on the left pane of the **Logic Simulator** window. Moreover, you will see an "i" besides an input signal and an "o" next to an output signal. The waveforms of the signals will be displayed in the order in which you added the signals from top to bottom. In this example, we choose to display all four inputs above the output. Therefore, DOOR is added first, followed by SEATBELT, IGNITION, HEADLIGHT, and finally, ALARM. When you finish adding all the signals in that order, your **Logic Stimulator** window should resemble Figure T2.5.

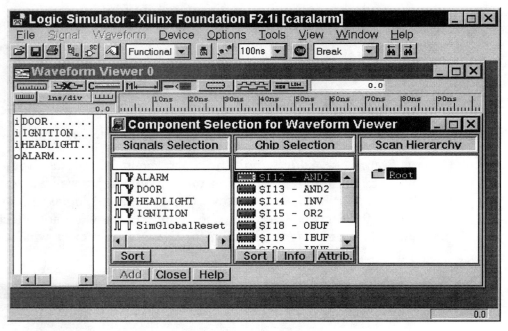

Figure T2.5 Logic Simulator window after step 6.

7) Click on the **Close** button in the **Component Selection for Waveform Viewer** window.

8) You have added all the signals that will participate in the simulation. We must now specify how we want the input signals to change in order to provide the complete result for the output. To do this, choose **Signal → Add Stimulators** from the menu bar of the **Logic Simulator** window (or click on the button). The **Stimulator Selection** window appears on top of the **Logic Simulator** window (Figure T2.6). In the **Stimulator Selection** window, the **Keyboard** pane provides signals that act like toggle switches. The **Clocks** pane will be used when a sequential circuit design is simulated. When we assign the stimulators from the buttons on the **Bc** (Binary counter) row, we will get all the combinations of the inputs. The 16 buttons represent the 16 outputs of a 16-bit binary counter. The rightmost button represents the least significant bit.

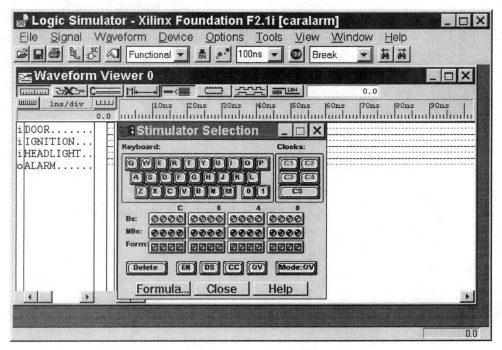

Figure T2.6 Logic Simulator window after step 8.

9)   We will assign stimulators to all three input signals beginning with the DOOR (The most significant bit in the truth table). First, highlight DOOR, and then click on the 3rd button from the right in the **Bc** row. You should see B2 indicated next to DOOR. Add stimulators B1 and B0 to IGINITION and HEADLIGHT in the same way. Click on the **Close** button to close the **Stimulator Selection** window.

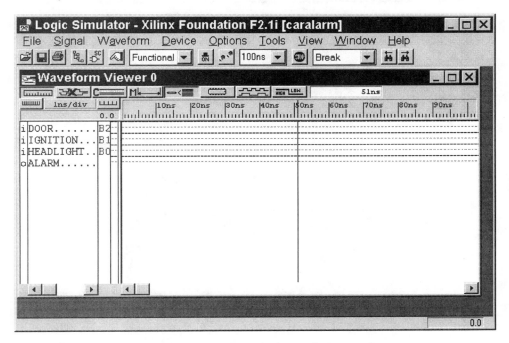

Figure T2.7 Logic Simulator window after step 9.

10) To start the simulation, choose **Options → Start Long Simulation** from the menu bar. Then click the **Start** button in the message window. After about 10 seconds, a set of waveforms similar to Figure T2.9 should appear.

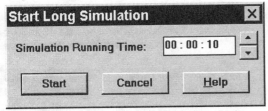

Figure T2.8 Message sub-window.

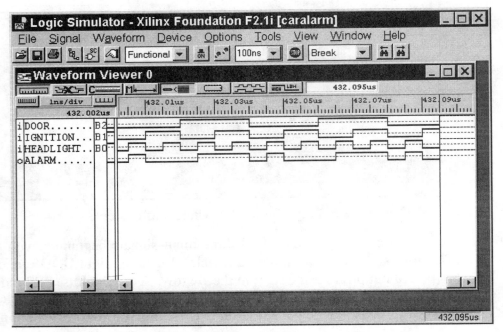

Figure T2.9 Logic Simulator window after step 10.

11) Click on the **Start Simulation** button ⬛. To view the waveforms clearly, you may have to scale the waveform by clicking on the **Zoom In** 🔲 or **Zoom Out** 🔲 buttons. Your **Logic Simulator** window should finally look similar to Figure T2.10.

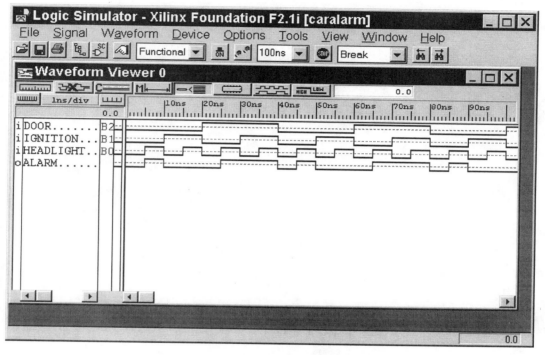

Figure T2.10 Logic Simulator window after step 11.

12) Compare the waveforms from step 10 to the result you obtained from Tutorial I or Table T1.2, they should agree with each other.

APPENDIX E: PARALLEL PORTS

It's possible that when you go to download a JEDEC file through the JTAG cable connected to your parallel port, you will get a message from the Xilinx Foundation software saying that it can't find the port. If that happens, don't panic. It could be you're using the wrong cable. Try a cable that contains all 25 wires. If that doesn't work, then most likely the solution is to reconfigure your parallel port.

The parallel port on a personal computer was originally designed as the interface for a printer. But because all PCs have one, and because it can be used for input as well as output, many external devices have been designed to connect to the parallel port.

Unlike the other common I/O interface, the serial port, the parallel port does not require the user to specify things like bit-rate and parity. However, over the years the implementation of parallel ports has gone through some evolution. There are at least four versions of the parallel port:

- the original "standard" parallel port, or SPP
- the simple bidirectional or PS/2 port
- the "enhanced" parallel port, or EPP
- the "extended capabilities" port, or ECP

While all versions of the parallel port will drive a printer, they are not all the same.

The SPP version can output data in 8-bit bytes, but can input data only in 4-bit nibbles. The SPP is not as fast as later versions. The PS/2 and later versions can both input and output data a byte at a time. The EPP version is faster; it can input or output data in one cycle of the bus clock instead of the four cycles required by earlier versions. The ECP version also requires one bus cycle, but also supports direct memory access (DMA) transfer of data.

Most new computers will emulate any of the four parallel port versions. If the JTAG programmer doesn't recognize your parallel port, select a different port mode. You can try going to the **Windows Control Panel** → **Systems Properties** → **Device Manager** → **Ports** and see if you can select another driver. Often, the version needs to be selected by going into the CMOS set-up page during the Windows boot-up. Go into set-up and select a different port mode.

If that doesn't cure the problem, then your target board may be defective.

281

APPENDIX F: DATA SHEETS

Copyright Notices

The following statement applies to all Xilinx data sheets in this appendix:

Figures based on or adapted from figures and text owned by Xilinx, Inc., courtesy of Xilinx, Inc. © 1998, 1999. All rights reserved.

The following statement applies to all XESS data sheets in this appendix:

Permission granted to reprint courtesy of XESS Corporation. Figures and text owned by XESS Corporation, © 1998, 1999, 2000. All rights reserved.

The following statement applies to all RSR Electronics data sheets in this appendix:

Printed by permission of the copyright holder, RSR Electronics, Inc., Avenel NJ. Figures and text owned by RSR Electronics, Inc., © 1999, 2000. All rights reserved.

XC9500 In-System Programmable CPLD Family

September 15, 1999 (Version 5.0)

Features

- High-performance
  - 5 ns pin-to-pin logic delays on all pins
  - f_{CNT} to 125 MHz
- Large density range
  - 36 to 288 macrocells with 800 to 6,400 usable gates
- 5 V in-system programmable
  - Endurance of 10,000 program/erase cycles
  - Program/erase over full commercial voltage and temperature range
- Enhanced pin-locking architecture
- Flexible 36V18 Function Block
  - 90 product terms drive any or all of 18 macrocells within Function Block
  - Global and product term clocks, output enables, set and reset signals
- Extensive IEEE Std 1149.1 boundary-scan (JTAG) support
- Programmable power reduction mode in each macrocell
- Slew rate control on individual outputs
- User programmable ground pin capability
- Extended pattern security features for design protection
- High-drive 24 mA outputs
- 3.3 V or 5 V I/O capability
- Advanced CMOS 5V FastFLASH technology
- Supports parallel programming of multiple XC9500 devices

Family Overview

The XC9500 CPLD family provides advanced in-system programming and test capabilities for high performance, general purpose logic integration. All devices are in-system programmable for a minimum of 10,000 program/erase cycles. Extensive IEEE 1149.1 (JTAG) boundary-scan support is also included on all family members.

As shown in Table 1, logic density of the XC9500 devices ranges from 800 to over 6,400 usable gates with 36 to 288 registers, respectively. Multiple package options and associated I/O capacity are shown in Table 2. The XC9500 family is fully pin-compatible allowing easy design migration across multiple density options in a given package footprint.

The XC9500 architectural features address the requirements of in-system programmability. Enhanced pin-locking capability avoids costly board rework. An expanded JTAG instruction set allows version control of programming patterns and in-system debugging. In-system programming throughout the full device operating range and a minimum of 10,000 program/erase cycles provide worry-free reconfigurations and system field upgrades.

Advanced system features include output slew rate control and user-programmable ground pins to help reduce system noise. I/Os may be configured for 3.3 V or 5 V operation. All outputs provide 24 mA drive.

Architecture Description

Each XC9500 device is a subsystem consisting of multiple Function Blocks (FBs) and I/O Blocks (IOBs) fully interconnected by the FastCONNECT switch matrix. The IOB provides buffering for device inputs and outputs. Each FB provides programmable logic capability with 36 inputs and 18 outputs. The FastCONNECT switch matrix connects all FB outputs and input signals to the FB inputs. For each FB, 12 to 18 outputs (depending on package pin-count) and associated output enable signals drive directly to the IOBs. See Figure 1.

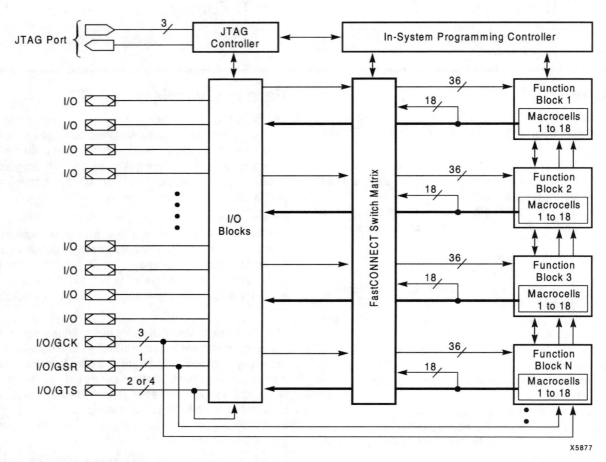

X5877

Figure 1: XC9500 Architecture

> **Note:** Function Block outputs (indicated by the bold line) drive the I/O Blocks directly.

Table 1: XC9500 Device Family

	XC9536	XC9572	XC95108	XC95144	XC95216	XC95288
Macrocells	36	72	108	144	216	288
Usable Gates	800	1,600	2,400	3,200	4,800	6,400
Registers	36	72	108	144	216	288
t_{PD} (ns)	5	7.5	7.5	7.5	10	10
t_{SU} (ns)	3.5	4.5	4.5	4.5	6.0	6.0
t_{CO} (ns)	4.0	4.5	4.5	4.5	6.0	6.0
f_{CNT} (MHz)	100	125	125	125	111.1	111.1
f_{SYSTEM} (MHz)	100	83.3	83.3	83.3	66.7	66.7

> **Note:** f_{CNT} = Operating frequency for 16-bit counters
> f_{SYSTEM} = Internal operating frequency for general purpose system designs spanning multiple FBs.

Table 2: Available Packages and Device I/O Pins (not including dedicated JTAG pins)

	XC9536	XC9572	XC95108	XC95144	XC95216	XC95288
44-Pin VQFP	34					
44-Pin PLCC	34	34				
48-Pin CSP	34					
84-Pin PLCC		69	69			
100-Pin TQFP		72	81	81		
100-Pin PQFP		72	81	81		
160-Pin PQFP			108	133	133	
208-Pin HQFP					166	168
352-Pin BGA					166	192

Function Block

Each Function Block, as shown in Figure 2, is comprised of 18 independent macrocells, each capable of implementing a combinatorial or registered function. The FB also receives global clock, output enable, and set/reset signals. The FB generates 18 outputs that drive the FastCONNECT switch matrix. These 18 outputs and their corresponding output enable signals also drive the IOB.

Logic within the FB is implemented using a sum-of-products representation. Thirty-six inputs provide 72 true and complement signals into the programmable AND-array to form 90 product terms. Any number of these product terms, up to the 90 available, can be allocated to each macrocell by the product term allocator.

Each FB (except for the XC9536) supports local feedback paths that allow any number of FB outputs to drive into its own programmable AND-array without going outside the FB. These paths are used for creating very fast counters and state machines where all state registers are within the same FB.

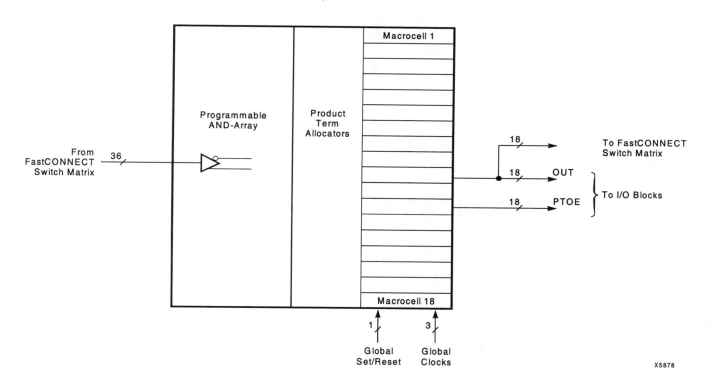

Figure 2: XC9500 Function Block

285

Macrocell

Each XC9500 macrocell may be individually configured for a combinatorial or registered function. The macrocell and associated FB logic is shown in Figure 3.

Five direct product terms from the AND-array are available for use as primary data inputs (to the OR and XOR gates) to implement combinatorial functions, or as control inputs including clock, set/reset, and output enable. The product

term allocator associated with each macrocell selects how the five direct terms are used.

The macrocell register can be configured as a D-type or T-type flip-flop, or it may be bypassed for combinatorial operation. Each register supports both asynchronous set and reset operations. During power-up, all user registers are initialized to the user-defined preload state (default to 0 if unspecified).

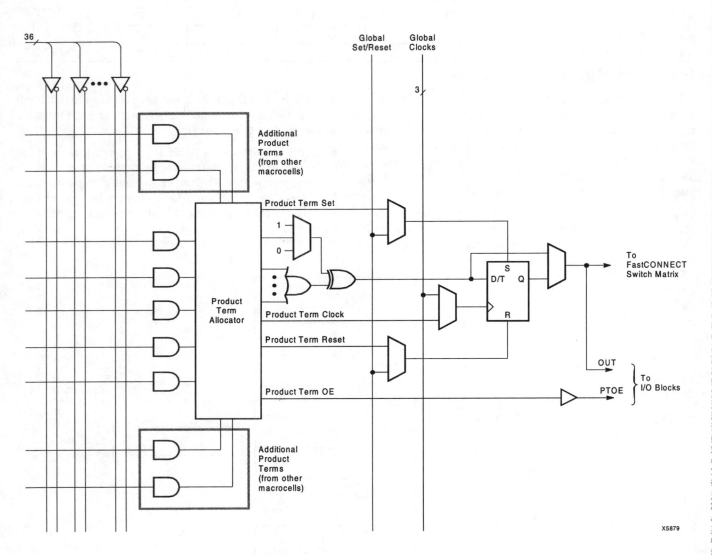

Figure 3:   XC9500 Marcocell Within Function Block

All global control signals are available to each individual macrocell, including clock, set/reset, and output enable signals. As shown in Figure 4, the macrocell register clock originates from either of three global clocks or a product term clock. Both true and complement polarities of a GCK pin can be used within the device. A GSR input is also provided to allow user registers to be set to a user-defined state.

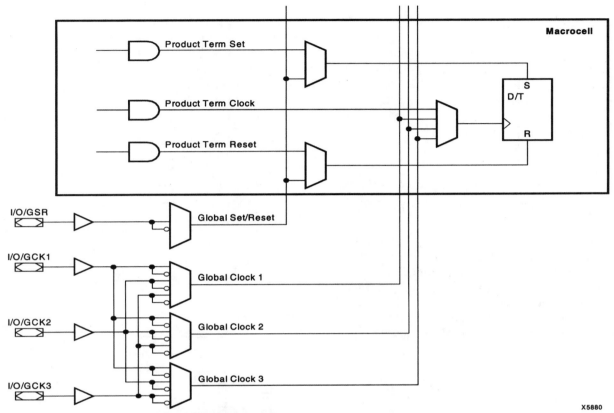

Figure 4: Macrocell Clock and Set/Reset Capability

X5880

Product Term Allocator

The product term allocator controls how the five direct product terms are assigned to each macrocell. For example, all five direct terms can drive the OR function as shown in Figure 5.

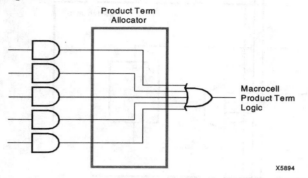

Figure 5: Macrocell Logic Using Direct Product Term

The product term allocator can re-assign other product terms within the FB to increase the logic capacity of a macrocell beyond five direct terms. Any macrocell requiring additional product terms can access uncommitted product terms in other macrocells within the FB. Up to 15 product terms can be available to a single macrocell with only a small incremental delay of t_{PTA}, as shown in Figure 6.

Note that the incremental delay affects only the product terms in other macrocells. The timing of the direct product terms is not changed.

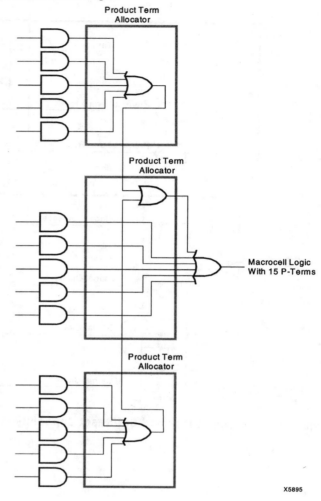

Figure 6: Product Term Allocation With 15 Product Terms

XC9500 In-System Programmable CPLD Family

The product term allocator can re-assign product terms from any macrocell within the FB by combining partial sums of products over several macrocells, as shown in Figure 7.

In this example, the incremental delay is only 2*t_{PTA}. All 90 product terms are available to any macrocell, with a maximum incremental delay of 8*t_{PTA}.

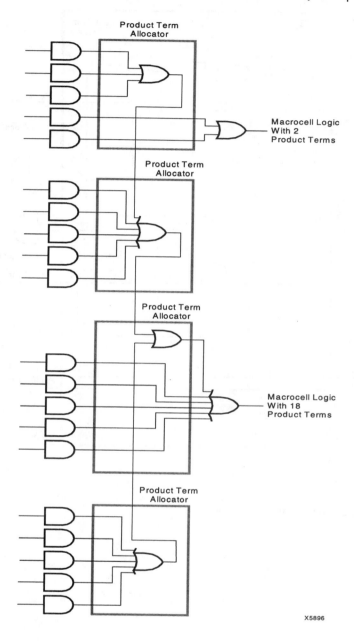

Figure 7: Product Term Allocation Over Several Macrocells

289

The internal logic of the product term allocator is shown in Figure 8.

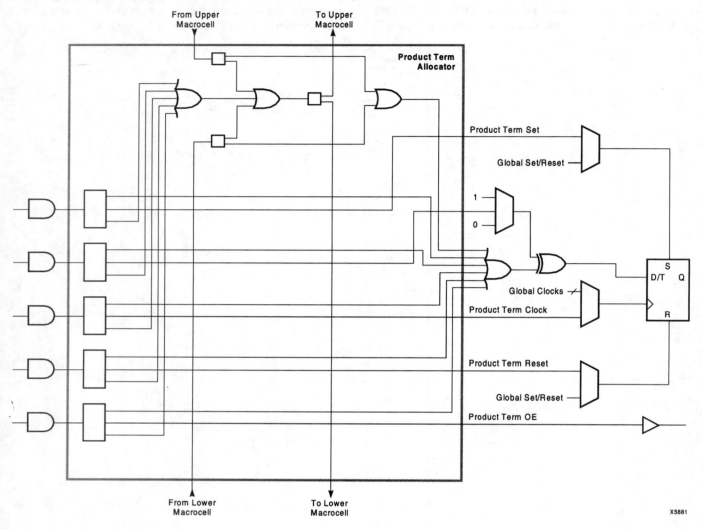

Figure 8: Product Term Allocator Logic

X5881

290

FastCONNECT Switch Matrix

The FastCONNECT switch matrix connects signals to the FB inputs, as shown in Figure 9. All IOB outputs (corresponding to user pin inputs) and all FB outputs drive the FastCONNECT matrix. Any of these (up to a FB fan-in limit of 36) may be selected, through user programming, to drive each FB with a uniform delay.

The FastCONNECT switch matrix is capable of combining multiple internal connections into a single wired-AND output before driving the destination FB. This provides additional logic capability and increases the effective logic fan-in of the destination FB without any additional timing delay. This capability is available for internal connections originating from FB outputs only. It is automatically invoked by the development software where applicable.

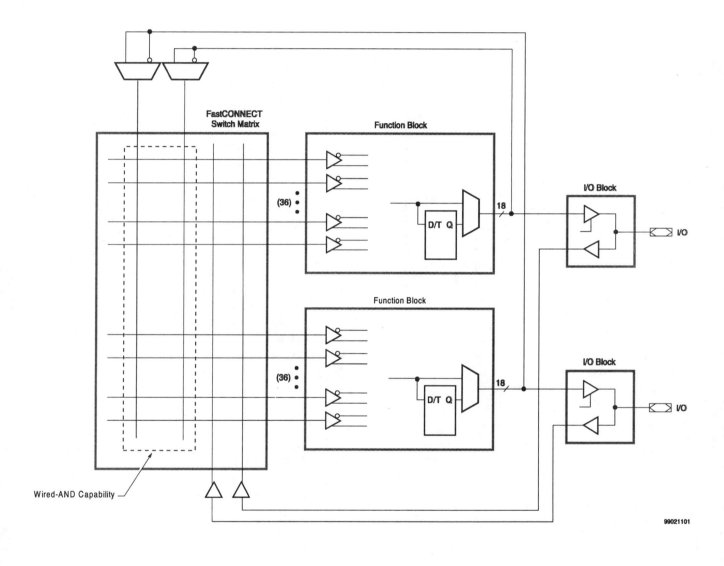

Figure 9: FastCONNECT Switch Matrix

I/O Block

The I/O Block (IOB) interfaces between the internal logic and the device user I/O pins. Each IOB includes an input buffer, output driver, output enable selection multiplexer, and user programmable ground control. See Figure 10 for details.

The input buffer is compatible with standard 5 V CMOS, 5 V TTL and 3.3 V signal levels. The input buffer uses the internal 5 V voltage supply (V_{CCINT}) to ensure that the input thresholds are constant and do not vary with the V_{CCIO} voltage.

The output enable may be generated from one of four options: a product term signal from the macrocell, any of the global OE signals, always "1", or always "0". There are two global output enables for devices with up to 144 macrocells, and four global output enables for devices with 180 or more macrocells. Both polarities of any of the global 3-state control (GTS) pins may be used within the device.

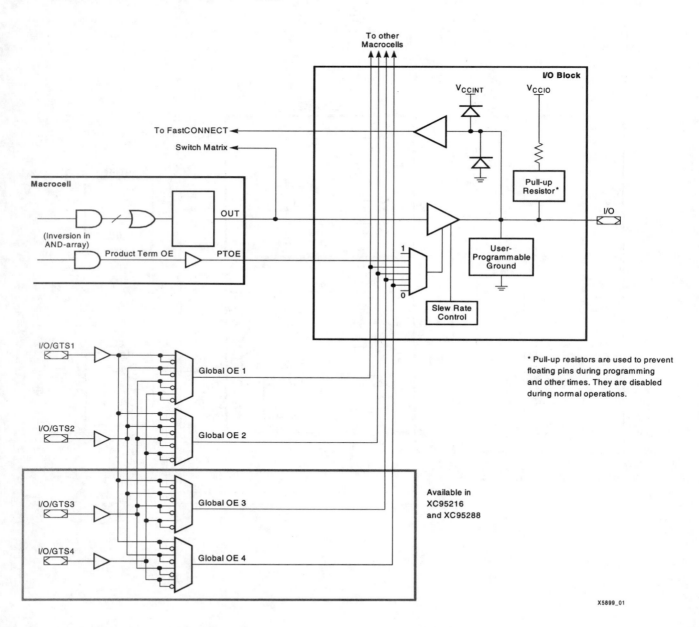

Figure 10: I/O Block and Output Enable Capability

Each output has independent slew rate control. Output edge rates may be slowed down to reduce system noise (with an additional time delay of t_{SLEW}) through programming. See Figure 11.

Each IOB provides user programmable ground pin capability. This allows device I/O pins to be configured as additional ground pins. By tying strategically located programmable ground pins to the external ground connection, system noise generated from large numbers of simultaneous switching outputs may be reduced.

A control pull-up resistor (typically 10K ohms) is attached to each device I/O pin to prevent them from floating when the device is not in normal user operation. This resistor is active during device programming mode and system power-up. It is also activated for an erased device. The resistor is deactivated during normal operation.

The output driver is capable of supplying 24 mA output drive. All output drivers in the device may be configured for either 5 V TTL levels or 3.3 V levels by connecting the device output voltage supply (V_{CCIO}) to a 5 V or 3.3 V

voltage supply. Figure 12 shows how the XC9500 device can be used in 5 V only and mixed 3.3 V/5 V systems.

Pin-Locking Capability

The capability to lock the user defined pin assignments during design changes depends on the ability of the architecture to adapt to unexpected changes. The XC9500 devices have architectural features that enhance the ability to accept design changes while maintaining the same pinout.

The XC9500 architecture provides maximum routing within the FastCONNECT switch matrix, and incorporates a flexible Function Block that allows block-wide allocation of available product terms. This provides a high level of confidence of maintaining both input and output pin assignments for unexpected design changes.

For extensive design changes requiring higher logic capacity than is available in the initially chosen device, the new design may be able to fit into a larger pin-compatible device using the same pin assignments. The same board may be used with a higher density device without the expense of board rework.

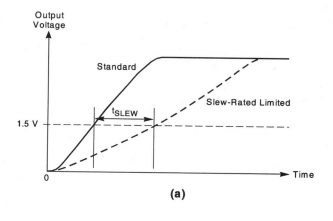
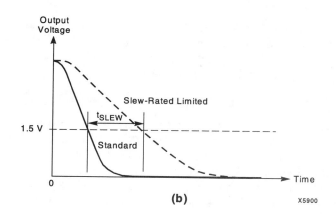

Figure 11:   Output Slew-Rate For (a) Rising and (b) Falling Outputs

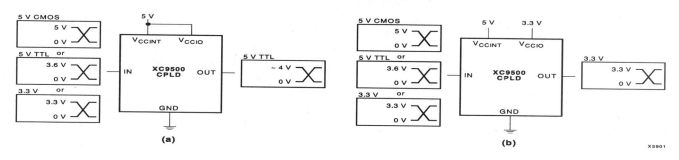

Figure 12:   XC9500 Devices in (a) 5 V Systems and (b) Mixed 3.3 V/5 V Systems

In-System Programming

XC9500 devices are programmed in-system via a standard 4-pin JTAG protocol, as shown in Figure 13. In-system programming offers quick and efficient design iterations and eliminates package handling. The Xilinx development system provides the programming data sequence using a Xilinx download cable, a third-party JTAG development system, JTAG-compatible board tester, or a simple microprocessor interface that emulates the JTAG instruction sequence.

All I/Os are 3-stated and pulled high by the IOB resistors during in-system programming. If a particular signal must remain low during this time, then a pulldown resistor may be added to the pin.

External Programming

XC9500 devices can also be programmed by the Xilinx HW130 device programmer as well as third-party programmers. This provides the added flexibility of using pre-programmed devices during manufacturing, with an in-system programmable option for future enhancements.

Endurance

All XC9500 CPLDs provide a minimum endurance level of 10,000 in-system program/erase cycles. Each device meets all functional, performance, and data retention specifications within this endurance limit.

IEEE 1149.1 Boundary-Scan (JTAG)

XC9500 devices fully support IEEE 1149.1 boundary-scan (JTAG). EXTEST, SAMPLE/PRELOAD, BYPASS, USER-CODE, INTEST, IDCODE, and HIGHZ instructions are supported in each device. For ISP operations, five additional instructions are added; the ISPEN, FERASE, FPGM, FVFY, and ISPEX instructions are fully compliant extensions of the 1149.1 instruction set.

The TMS and TCK pins have dedicated pull-up resistors as specified by the IEEE 1149.1 standard.

Boundary Scan Description Language (BSDL) files for the XC9500 are included in the development system and are available on the Xilinx FTP site.

Design Security

XC9500 devices incorporate advanced data security features which fully protect the programming data against unauthorized reading or inadvertent device erasure/reprogramming. Table 3 shows the four different security settings available.

The read security bits can be set by the user to prevent the internal programming pattern from being read or copied. When set, they also inhibit further program operations but allow device erase. Erasing the entire device is the only way to reset the read security bit.

The write security bits provide added protection against accidental device erasure or reprogramming when the JTAG pins are subject to noise, such as during system power-up. Once set, the write-protection may be deactivated when the device needs to be reprogrammed with a valid pattern.

Table 3: Data Security Options

		Read Security	
		Default	**Set**
Write Security	**Default**	Read Allowed Program/Erase Allowed	Read Inhibited Program Inhibited/Erase Allowed
	Set	Read Allowed Program/Erase Inhibited	Read Inhibited Program/Erase Inhibited

X5905

294

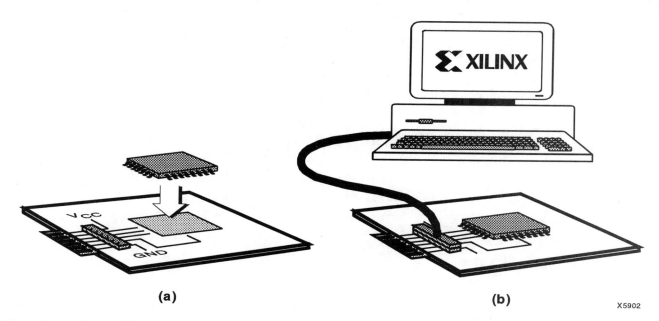

Figure 13: In-System Programming Operation (a) Solder Device to PCB and (b) Program Using Download Cable

Low Power Mode

All XC9500 devices offer a low-power mode for individual macrocells or across all macrocells. This feature allows the device power to be significantly reduced.

Each individual macrocell may be programmed in low-power mode by the user. Performance-critical parts of the application can remain in standard power mode, while other parts of the application may be programmed for low-power operation to reduce the overall power dissipation. Macrocells programmed for low-power mode incur additional delay (t_{LP}) in pin-to-pin combinatorial delay as well as register setup time. Product term clock to output and product term output enable delays are unaffected by the macrocell power-setting.

Timing Model

The uniformity of the XC9500 architecture allows a simplified timing model for the entire device. The basic timing model, shown in Figure 14, is valid for macrocell functions that use the direct product terms only, with standard power setting, and standard slew rate setting. Table 4 shows how each of the key timing parameters is affected by the product term allocator (if needed), low-power setting, and slew-limited setting.

The product term allocation time depends on the logic span of the macrocell function, which is defined as one less than the maximum number of allocators in the product term path. If only direct product terms are used, then the logic span is 0. The example in Figure 6 shows that up to 15 product terms are available with a span of 1. In the case of Figure 7, the 18 product term function has a span of 2.

Detailed timing information may be derived from the full timing model shown in Figure 15. The values and explanations for each parameter are given in the individual device data sheets.

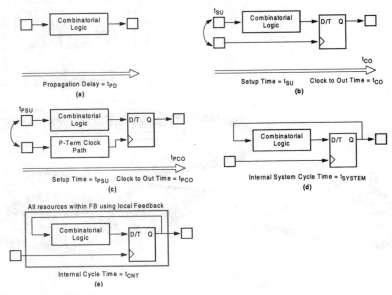

Figure 14: Basic Timing Model

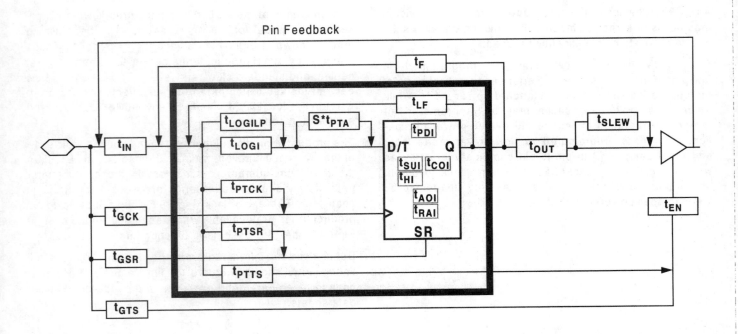

Figure 15: Detailed Timing Model

Power-Up Characteristics

The XC9500 devices are well behaved under all operating conditions. During power-up each XC9500 device employs internal circuitry which keeps the device in the quiescent state until the V_{CCINT} supply voltage is at a safe level (approximately 3.8 V). During this time, all device pins and JTAG pins are disabled and all device outputs are disabled with the IOB pull-up resistors (~ 10K ohms) enabled, as shown in Table 5. When the supply voltage reaches a safe level, all user registers become initialized (typically within 100 µs for 9536 - 95144, 200 µs for 95216 and 300 µs for 95288), and the device is immediately available for operation, as shown in Figure 16.

If the device is in the erased state (before any user pattern is programmed), the device outputs remain disabled with the IOB pull-up resistors enabled. The JTAG pins are enabled to allow the device to be programmed at any time.

If the device is programmed, the device inputs and outputs take on their configured states for normal operation. The JTAG pins are enabled to allow device erasure or boundary-scan tests at any time.

Development System Support

The XC9500 CPLD family is fully supported by the development systems available from Xilinx and the Xilinx Alliance Program vendors.

The designer can create the design using ABEL, schematics, equations, VHDL, or Verilog in a variety of software front-end tools. The development system can be used to implement the design and generate a JEDEC bitmap which can be used to program the XC9500 device. Each development system includes JTAG download software that can be used to program the devices via the standard JTAG interface and a download cable.

FastFLASH Technology

An advanced CMOS Flash process is used to fabricate all XC9500 devices. Specifically developed for Xilinx in-system programmable CPLDs, the FastFLASH process provides high performance logic capability, fast programming times, and endurance of 10,000 program/erase cycles.

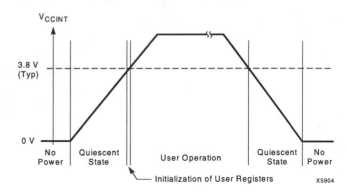

Figure 16: Device Behavior During Power-up

Table 4: Timing Model Parameters

Description	Parameter	Product Term Allocator[1]	Macrocell Low-Power Setting	Output Slew-Limited Setting
Propagation Delay	t_{PD}	+ t_{PTA} * S	+ t_{LP}	+ t_{SLEW}
Global Clock Setup Time	t_{SU}	+ t_{PTA} * S	+ t_{LP}	–
Global Clock-to-output	t_{CO}	–	–	+ t_{SLEW}
Product Term Clock Setup Time	t_{PSU}	+ t_{PTA} * S	+ t_{LP}	–
Product Term Clock-to-output	t_{PCO}	–	–	+ t_{SLEW}
Internal System Cycle Period	t_{SYSTEM}	+ t_{PTA} * S	+ t_{LP}	–

Note: 1. S = the logic span of the function, as defined in the text.

Table 5: XC9500 Device Characteristics

Device Circuitry	Quiescent State	Erased Device Operation	Valid User Operation
IOB Pull-up Resistors	Enabled	Enabled	Disabled
Device Outputs	Disabled	Disabled	As Configured
Device Inputs and Clocks	Disabled	Disabled	As Configured
Function Block	Disabled	Disabled	As Configured
JTAG Controller	Disabled	Enabled	Enabled

Revision History

Version	Date	Revision
3.0	12/14/98	Revised datasheet to reflect new AC characteristics and Internal Timing Parmeters.
4.0	2/10/99	Corrected Figure 3
5.0	9/15/99	Added -10 speed grade to 95288

XC95108 In-System Programmable CPLD

December 4, 1998 (Version 3.0)

Product Specification

Features

- 7.5 ns pin-to-pin logic delays on all pins
- f_{CNT} to 125 MHz
- 108 macrocells with 2400 usable gates
- Up to 108 user I/O pins
- 5 V in-system programmable (ISP)
  - Endurance of 10,000 program/erase cycles
  - Program/erase over full commercial voltage and temperature range
- Enhanced pin-locking architecture
- Flexible 36V18 Function Block
  - 90 product terms drive any or all of 18 macrocells within Function Block
  - Global and product term clocks, output enables, set and reset signals
- Extensive IEEE Std 1149.1 boundary-scan (JTAG) support
- Programmable power reduction mode in each macrocell
- Slew rate control on individual outputs
- User programmable ground pin capability
- Extended pattern security features for design protection
- High-drive 24 mA outputs
- 3.3 V or 5 V I/O capability
- Advanced CMOS 5V FastFLASH technology
- Supports parallel programming of more than one XC9500 concurrently
- Available in 84-pin PLCC, 100-pin PQFP, 100-pin TQFP and 160-pin PQFP packages

Description

The XC95108 is a high-performance CPLD providing advanced in-system programming and test capabilities for general purpose logic integration. It is comprised of six 36V18 Function Blocks, providing 2,400 usable gates with propagation delays of 7.5 ns. See Figure 2 for the architecture overview.

Power Management

Power dissipation can be reduced in the XC95108 by configuring macrocells to standard or low-power modes of operation. Unused macrocells are turned off to minimize power dissipation.

Operating current for each design can be approximated for specific operating conditions using the following equation:

$$I_{CC} \text{ (mA)} =$$

$$MC_{HP} (1.7) + MC_{LP} (0.9) + MC (0.006 \text{ mA/MHz}) f$$

Where:

MC_{HP} = Macrocells in high-performance mode

MC_{LP} = Macrocells in low-power mode

MC = Total number of macrocells used

f = Clock frequency (MHz)

Figure 1 shows a typical calculation for the XC95108 device.

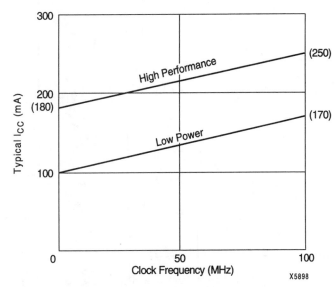

Figure 1: Typical I_{CC} vs. Frequency for XC95108

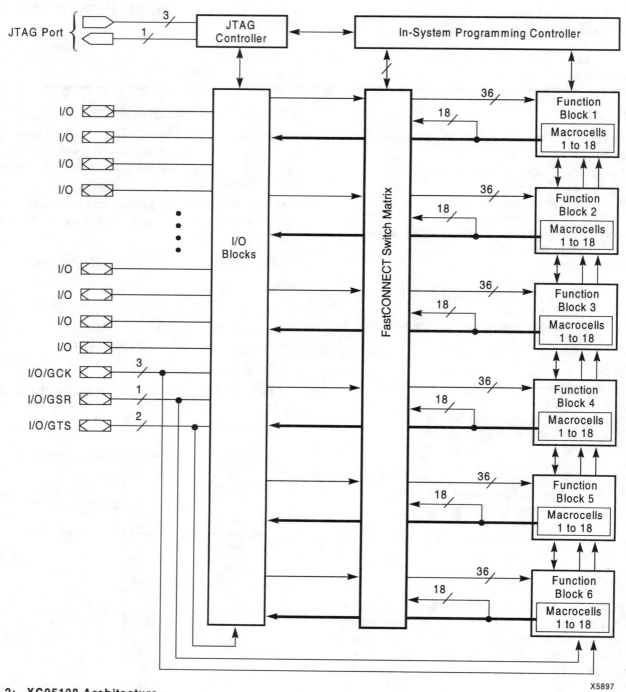

Figure 2: XC95108 Architecture

X5897

Note: Function Block outputs (indicated by the bold line) drive the I/O Blocks directly

Absolute Maximum Ratings

Symbol	Parameter	Value	Units
V_{CC}	Supply voltage relative to GND	-0.5 to 7.0	V
V_{IN}	DC input voltage relative to GND	-0.5 to V_{CC} + 0.5	V
V_{TS}	Voltage applied to 3-state output with respect to GND	-0.5 to V_{CC} + 0.5	V
T_{STG}	Storage temperature	-65 to +150	°C
T_{SOL}	Max soldering temperature (10 s @ 1/16 in = 1.5 mm)	+260	°C

Warning: Stresses beyond those listed under Absolute Maximum Ratings may cause permanent damage to the device. These are stress ratings only, and functional operation of the device at these or any other conditions beyond those listed under Recommended Operating Conditions is not implied. Exposure to Absolute Maximum Rating conditions for extended periods of time may affect device reliability.

Recommended Operation Conditions[1]

Symbol	Parameter	Min	Max	Units
V_{CCINT}	Supply voltage for internal logic and input buffer	4.75 (4.5)	5.25 (5.5)	V
V_{CCIO}	Supply voltage for output drivers for 5 V operation	4.75 (4.5)	5.25 (5.5)	V
	Supply voltage for output drivers for 3.3 V operation	3.0	3.6	V
V_{IL}	Low-level input voltage	0	0.80	V
V_{IH}	High-level input voltage	2.0	V_{CCINT} +0.5	V
V_O	Output voltage	0	V_{CCIO}	V

Note: 1. Numbers in parenthesis are for industrial-temperature range versions.

Endurance Characteristics

Symbol	Parameter	Min	Max	Units
t_{DR}	Data Retention	20	-	Years
N_{PE}	Program/Erase Cycles	10,000	-	Cycles

DC Characteristics Over Recommended Operating Conditions

Symbol	Parameter	Test Conditions	Min	Max	Units
V_{OH}	Output high voltage for 5 V operation	I_{OH} = -4.0 mA V_{CC} = Min	2.4		V
	Output high voltage for 3.3 V operation	I_{OH} = -3.2 mA V_{CC} = Min	2.4		V
V_{OL}	Output low voltage for 5 V operation	I_{OL} = 24 mA V_{CC} = Min		0.5	V
	Output low voltage for 3.3 V operation	I_{OL} = 10 mA V_{CC} = Min		0.4	V
I_{IL}	Input leakage current	V_{CC} = Max V_{IN} = GND or V_{CC}		±10.0	µA
I_{IH}	I/O high-Z leakage current	V_{CC} = Max V_{IN} = GND or V_{CC}		±10.0	µA
C_{IN}	I/O capacitance	V_{IN} = GND f = 1.0 MHz		10.0	pF
I_{CC}	Operating Supply Current (low power mode, active)	V_I = GND, No load f = 1.0 MHz	100 (Typ)		ma

AC Characteristics

Symbol	Parameter	XC95108-7		XC95108-10		XC95108-15		XC95108-20		Units
		Min	Max	Min	Max	Min	Max	Min	Max	
t_{PD}	I/O to output valid		7.5		10.0		15.0		20.0	ns
t_{SU}	I/O setup time before GCK	4.5		6.0		8.0		10.0		ns
t_H	I/O hold time after GCK	0.0		0.0		0.0		0.0		ns
t_{CO}	GCK to output valid		4.5		6.0		8.0		10.0	ns
f_{CNT}[1]	16-bit counter frequency	125.0		111.1		95.2		83.3		MHz
f_{SYSTEM}[2]	Multiple FB internal operating frequency	83.3		66.7		55.6		50.0		MHz
t_{PSU}	I/O setup time before p-term clock input	0.5		2.0		4.0		4.0		ns
t_{PH}	I/O hold time after p-term clock input	4.0		4.0		4.0		6.0		ns
t_{PCO}	P-term clock to output valid		8.5		10.0		12.0		16.0	ns
t_{OE}	GTS to output valid		5.5		6.0		11.0		16.0	ns
t_{OD}	GTS to output disable		5.5		6.0		11.0		16.0	ns
t_{POE}	Product term OE to output enabled		9.5		10.0		14.0		18.0	ns
t_{POD}	Product term OE to output disabled		9.5		10.0		14.0		18.0	ns
t_{WLH}	GCK pulse width (High or Low)	4.0		4.5		5.5		5.5		ns

Note: 1. f_{CNT} is the fastest 16-bit counter frequency available, using the local feedback when applicable.
f_{CNT} is also the Export Control Maximum flip-flop toggle rate, f_{TOG}.
2. f_{SYSTEM} is the internal operating frequency for general purpose system designs spanning multiple FBs.

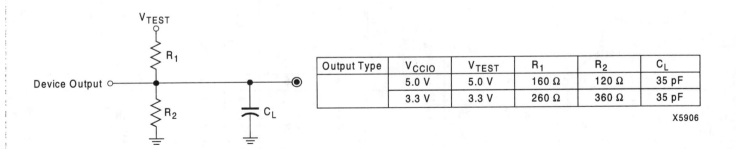

Output Type	V_CCIO	V_TEST	R_1	R_2	C_L
	5.0 V	5.0 V	160 Ω	120 Ω	35 pF
	3.3 V	3.3 V	260 Ω	360 Ω	35 pF

X5906

Figure 3: AC Load Circuit

Internal Timing Parameters

Symbol	Parameter	XC95108-7		XC95108-10		XC95108-15		XC95108-20		Units
		Min	Max	Min	Max	Min	Max	Min	Max	
Buffer Delays										
t_{IN}	Input buffer delay		2.5		3.5		4.5		6.5	ns
t_{GCK}	GCK buffer delay		1.5		2.5		3.0		3.0	ns
t_{GSR}	GSR buffer delay		4.5		6.0		7.5		9.5	ns
t_{GTS}	GTS buffer delay		5.5		6.0		11.0		16.0	ns
t_{OUT}	Output buffer delay		2.5		3.0		4.5		6.5	ns
t_{EN}	Output buffer enable/disable delay		0.0		0.0		0.0		0.0	ns
Product Term Control Delays										
t_{PTCK}	Product term clock delay		3.0		3.0		2.5		2.5	ns
t_{PTSR}	Product term set/reset delay		2.0		2.5		3.0		3.0	ns
t_{PTTS}	Product term 3-state delay		4.5		3.5		5.0		5.0	ns
Internal Register and Combinatorial delays										
t_{PDI}	Combinatorial logic propagation delay		0.5		1.0		3.0		4.0	ns
t_{SUI}	Register setup time	1.5		2.5		3.5		3.5		ns
t_{HI}	Register hold time	3.0		3.5		4.5		6.5		ns
t_{COI}	Register clock to output valid time		0.5		0.5		0.5		0.5	ns
t_{AOI}	Register async. S/R to output delay		6.5		7.0		8.0		8.0	ns
t_{RAI}	Register async. S/R recovery before clock	7.5		10.0		10.0		10.0		ns
t_{LOGI}	Internal logic delay		2.0		2.5		3.0		3.0	ns
t_{LOGILP}	Internal low power logic delay		10.0		11.0		11.5		11.5	ns
Feedback Delays										
t_F	FastCONNECT matrix feedback delay		8.0		9.5		11.0		13.0	ns
t_{LF}	Function Block local feeback delay		4.0		3.5		3.5		5.0	ns
Time Adders										
t_{PTA}[3]	Incremental Product Term Allocator delay		1.0		1.0		1.0		1.5	ns
t_{SLEW}	Slew-rate limited delay		4.0		4.5		5.0		5.5	ns

Note: 3. t_{PTA} is multiplied by the span of the function as defined in the family data sheet.

XC95108 I/O Pins

Function Block	Macrocell	PC84	PQ100	TQ100	PQ160	BScan Order	Notes	Function Block	Macrocell	PC84	PQ100	TQ100	PQ160	BScan Order	Notes
1	1	–	–	–	25	321		3	1	–	–	–	45	213	
1	2	1	15	13	21	318		3	2	14	31	29	47	210	
1	3	2	16	14	22	315		3	3	15	32	30	49	207	
1	4	–	21	19	29	312		3	4	–	36	34	57	204	
1	5	3	17	15	23	309		3	5	17	34	32	54	201	
1	6	4	18	16	24	306		3	6	18	35	33	56	198	
1	7	–	–	–	27	303		3	7	–	–	–	50	195	
1	8	5	19	17	26	300		3	8	19	37	35	58	192	
1	9	6	20	18	28	297		3	9	20	38	36	59	189	
1	10	–	26	24	36	294		3	10	–	45	43	69	186	
1	11	7	22	20	30	291		3	11	21	39	37	60	183	
1	12	9	24	22	33	288	[1]	3	12	23	41	39	62	180	
1	13	–	–	–	34	285		3	13	–	–	–	52	177	
1	14	10	25	23	35	282	[1]	3	14	24	42	40	63	174	
1	15	11	27	25	37	279		3	15	25	43	41	64	171	
1	16	12	29	27	42	276	[1]	3	16	26	44	42	68	168	
1	17	13	30	28	44	273		3	17	31	51	49	77	165	
1	18	–	–	–	43	270		3	18	–	–	–	74	162	
2	1	–	–	–	158	267		4	1	–	–	–	123	159	
2	2	71	98	96	154	264		4	2	57	83	81	134	156	
2	3	72	99	97	156	261		4	3	58	84	82	135	153	
2	4	–	4	2	4	258		4	4	–	82	80	133	150	
2	5	74	1	99	159	255	[1]	4	5	61	87	85	138	147	
2	6	75	3	1	2	252		4	6	62	88	86	139	144	
2	7	–	–	–	9	249		4	7	–	–	–	128	141	
2	8	76	5	3	6	246	[1]	4	8	63	89	87	140	138	
2	9	77	6	4	8	243	[1]	4	9	65	91	89	142	135	
2	10	–	9	7	12	240		4	10	–	–	–	147	132	
2	11	79	8	6	11	237		4	11	66	92	90	143	129	
2	12	80	10	8	13	234		4	12	67	93	91	144	126	
2	13	–	–	–	14	231		4	13	–	–	–	153	123	
2	14	81	11	9	15	228		4	14	68	95	93	146	120	
2	15	82	12	10	17	225		4	15	69	96	94	148	117	
2	16	83	13	11	18	222		4	16	–	94	92	145	114	
2	17	84	14	12	19	219		4	17	70	97	95	152	111	
2	18	–	–	–	16	216		4	18	–	–	–	155	108	

Notes: [1] Global control pin

XC95108 I/O Pins (continued)

Function Block	Macrocell	PC84	PQ100	TQ100	PQ160	BScan Order	Notes	Function Block	Macrocell	PC84	PQ100	TQ100	PQ160	BScan Order	Notes
5	1	–	–	–	76	105		6	1	–	–	–	91	51	
5	2	32	52	50	79	102		6	2	45	67	65	103	48	
5	3	33	54	52	82	99		6	3	46	68	66	104	45	
5	4	–	48	46	72	96		6	4	–	75	73	116	42	
5	5	34	55	53	86	93		6	5	47	69	67	106	39	
5	6	35	56	54	88	90		6	6	48	70	68	108	36	
5	7	–	–	–	78	87		6	7	–	–	–	105	33	
5	8	36	57	55	90	84		6	8	50	72	70	111	30	
5	9	37	58	56	92	81		6	9	51	73	71	113	27	
5	10	–	–	–	84	78		6	10	–	–	–	107	24	
5	11	39	60	58	95	75		6	11	52	74	72	115	21	
5	12	40	62	60	97	72		6	12	53	76	74	117	18	
5	13	–	–	–	87	69		6	13	–	–	–	112	15	
5	14	41	63	61	98	66		6	14	54	78	76	122	12	
5	15	43	65	63	101	63		6	15	55	79	77	124	9	
5	16	–	61	59	96	60		6	16	–	81	79	129	6	
5	17	44	66	64	102	57		6	17	56	80	78	126	3	
5	18	–	–	–	89	54		6	18	–	–	–	114	0	

XC95108 Global, JTAG and Power Pins

Pin Type	PC84	PQ100	TQ100	PQ160
I/O/GCK1	9	24	22	33
I/O/GCK2	10	25	23	35
I/O/GCK3	12	29	27	42
I/O/GTS1	76	5	3	6
I/O/GTS2	77	6	4	8
I/O/GSR	74	1	99	159
TCK	30	50	48	75
TDI	28	47	45	71
TDO	59	85	83	136
TMS	29	49	47	73
V_{CCINT} 5 V	38,73,78	7,59,100	5,57,98	10,46,94,157
V_{CCIO} 3.3 V/5 V	22,64	28,40,53,90	26,38,51,88	1,41,61,81,121,141
GND	8,16,27,42,49,60	2,23,33,46,64,71,77,86	100,21,31,44,62,69,75,84	20,31,40,51,70,80,99
GND	–	–	–	100,110,120,127,137
GND	–	–	–	160
No connects	–	–	–	3,5,7,32,38,39,48,53,55,65,66,67,83,85,93,109,118,119,125,130,131,132,149,150,151

Ordering Information

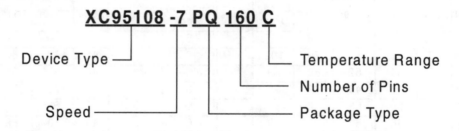

Speed Options

- 20 20 ns pin-to-pin delay
- -15 15 ns pin-to-pin delay
- -10 10 ns pin-to-pin delay
- -7   7 ns pin-to-pin delay

Packaging Options

PC84   84-Pin Plastic Leaded Chip Carrier (PLCC)
PQ100 100-Pin Plastic Quad Flat Pack (PQFP)
TQ100 100-Pin Very Thin Quad Flat Pack (TQFP)
PQ160 160-Pin Plastic Quad Flat Pack (PQFP)

Temperature Options

C   Commercial 0°C to +70°C
I   Industrial   −40°C to +85°C

Component Availability

Pins		84	100		160
Type		Plastic PLCC	Plastic PQFP	Plastic TQFP	Plastic PQFP
Code		PC84	PQ100	TQ100	PQ160
XC95108	−20	C(I)	C(I)	C(I)	C(I)
	−15	C(I)	C(I)	C(I)	C(I)
	−10	C(I)	C(I)	C(I)	C(I)
	−7	C(I)	C(I)	C(I)	C(I)

C = Commercial = 0° to +70°C I = Industrial = −40° to +85°C

Revision Control

Date	Revision
12/04/98	Update AC Characteristics and Internal Parameters

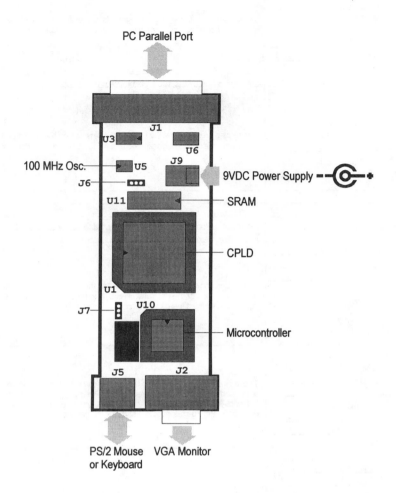

PC Parallel Port

J1

U3

U6

100 MHz Osc.　　　U5　　　J9

J6　　　　　　　　　　　9VDC Power Supply

U11　　　　　　　　　SRAM

CPLD

U1

J7　　　U10

Microcontroller

J5　　　　J2

PS/2 Mouse　VGA Monitor
or Keyboard

• Figure 2: Arrangement of components on the XS95 Board.

Connecting a PC to Your XS95 Board

The 6' cable included with your XS95 Board connects it to a PC. One end of the cable attaches to the parallel port on the PC and the other connects to the female DB-25 connector (J1) at the top of the XS95 Board as shown in Figure 1.

Connecting a VGA Monitor to Your XS95 Board

You can display images on a VGA monitor by connecting it to the 15-pin J2 connector at the bottom of your XS95 Board (see Figure 1). You will have to download a VGA driver circuit to your XS95 Board to actually display an image. You can find an example VGA driver at http://www.xess.com.

XS95 Pin	Connects to...	Description
21	S0.BLUE0	These pins drive the individual segments of the LED display (S0-S6 and DP). They also drive the color, horizontal, and vertical sync signals for a VGA monitor.
23	S1.BLUE1	
19	S2.GREEN0	
17	S3.GREEN1	
18	S4.RED0	
14	S5.RED1	
15	S6.HSYNCB	
24	DP.VSYNCB	
9	CLK	An input driven by the 100 MHz programmable oscillator.
46	PC_D0	These pins are driven by the data output pins of the PC parallel port. Clocking signals can only be reliably applied through pins 46 and 47 since these have additional hysterisis circuitry.
47	PC_D1	
48	PC_D2	
50	PC_D3	
51	PC_D4	
52	PC_D5	
81	PC_D6	
80	PC_D7	
10	XTAL1	Pin that drives the uC clock input
45	RST	Pin that drives the uC reset input
20	ALEB	Pin that monitors the uC address latch enable
13	PSENB	Pin that monitors the uC program store enable
6	P1.0.PC_C0	These pins connect to the pins of Port 1 of the uC. Some of the pins are also connected to the status input pins of the PC parallel port. The P1.0 port pin of the uC is also connected to the C0 control output from the parallel port.
7	P1.1	
11	P1.2	
5	P1.3	
72	P1.4.PC_S4	
71	P1.5.PC_S3	
66	P1.6.PC_S5	
67	P1.7	
31	P3.0(RXD)	These pins connect to the pins of Port 3 of the uC. The uC has specialized functions for each of the port pins indicated in parentheses. Pin 63 connects to the data write pin of the uC and the write-enable pin of the SRAM. Pins 26 and 70 connect to the clock and data lines of the PS/2 port. Pin 70 connects to a status input pin of the PC parallel port.
70	P3.1(TXD).PC_S6.KB_DATA	
69	P3.2(INTB0)	
68	P3.3(INTB1)	
26	P3.4(T0).KB_CLK	
33	P3.5(T1)	
63	P3.6(WRB).WEB	
32	P3.7(RDB)	
44	P0.0(AD0).D0	These pins connect to Port 0 of the uC which is also a multiplexed address/data port. These pins also connect to the data pins of the SRAM.
43	P0.1(AD1).D1	
41	P0.2(AD2).D2	
40	P0.3(AD3).D3	
39	P0.4(AD4).D4	
37	P0.5(AD5).D5	
36	P0.6(AD6).D6	
35	P0.7(AD7).D7	
58	P2.0(A8).A8	These pins connect to Port 2 of the uC which also outputs the upper address byte. These pins also connect to the upper address bits of the SRAM. Pins 34 and 74 are connected to the 128 KB SRAM address pins only on the XS95+ Board. Pins 34 and 74 do not connect to the 32 KB SRAM on the XS95 Board.
56	P2.0(A9).A9	
54	P2.0(A10).A10	
55	P2.0(A11).A11	
53	P2.0(A12).A12	
57	P2.0(A13).A13	
61	P2.0(A14).A14	
34	P2.0(A15).A15	
74	A16	
75	A0	These pins drive the 8 lower address bits of the SRAM.
79	A1	
82	A2	
84	A3	
1	A4	
3	A5	
83	A6	
2	A7	
62	OEB	Pin that drives the SRAM output enable.
65	CEB	Pin that drives the SRAM chip enable.
4	FREE0	These pins are not connected to other devices and can be used as general purpose I/O.
12	FREE1	
25	FREE2	
76	FREE3	
77	FREE4	

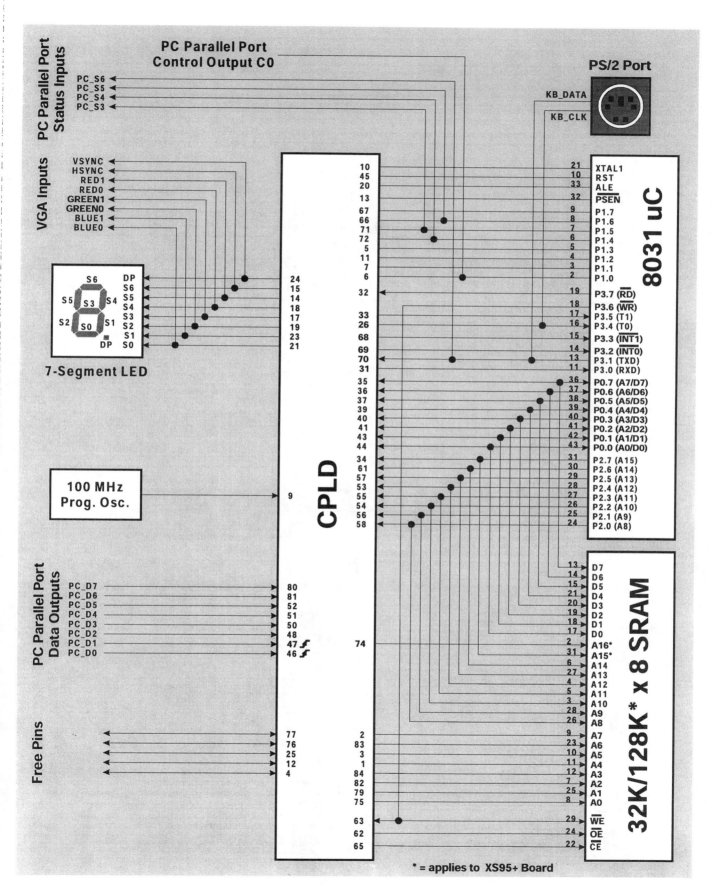

* = applies to XS95+ Board

309

XS95 and XS95+ Board V1.3 Schematic

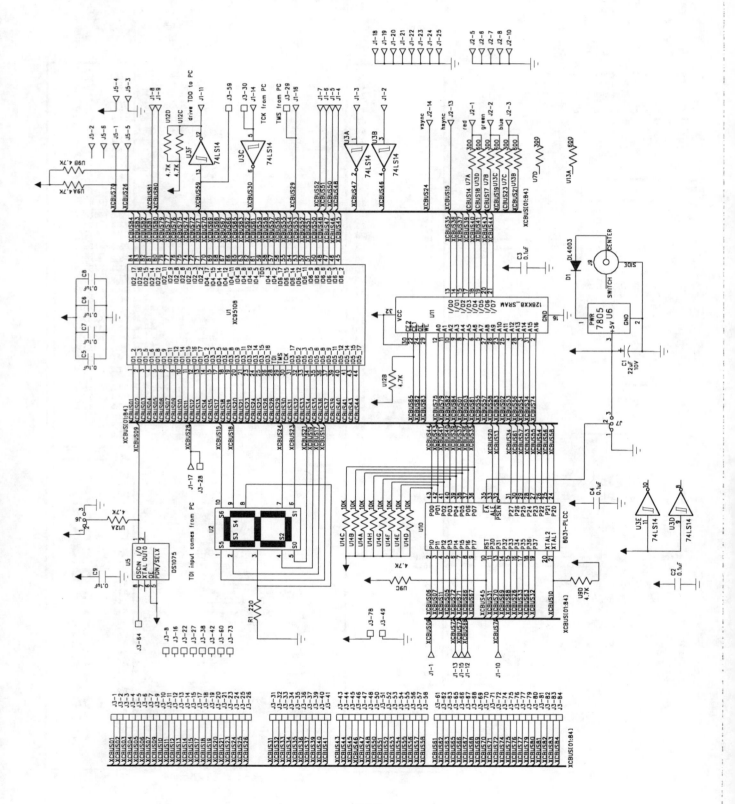

310

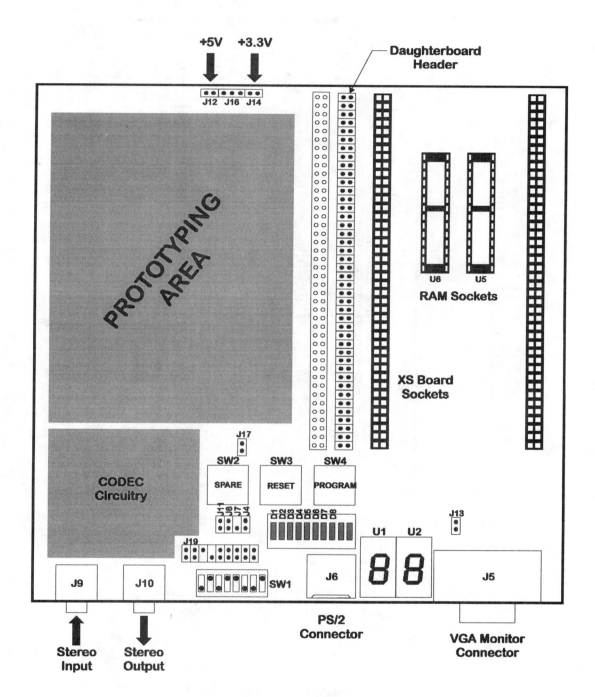

- **Figure 1:** XStend Board layout.

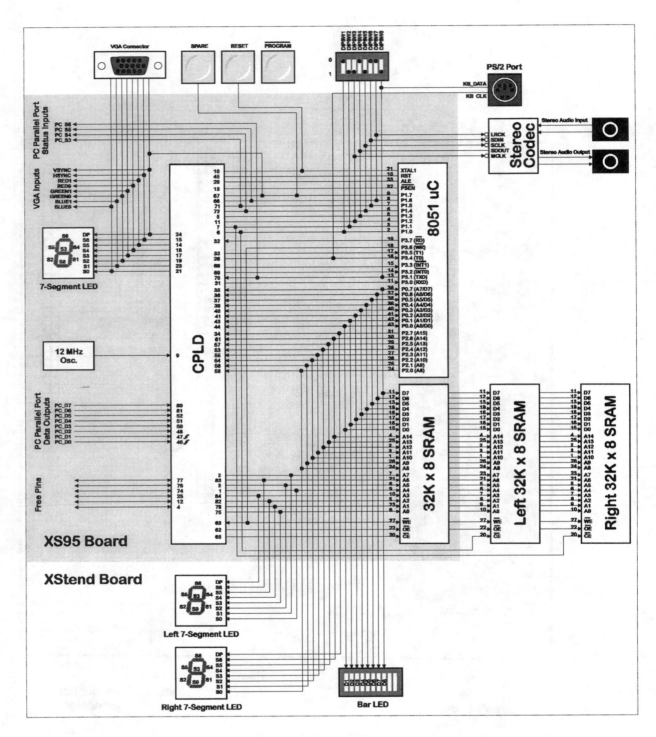

• **Figure 6:** Programmer's model of the XS95/XStend Board combination.

• **Table 5:** Connections between the XS95 Board and the XStend Board resources.

XS95 Pins (J2)	Power/ GND	DIP Switch	Push-buttons	LEDs	VGA Interface	PS/2 Interface	RAMs	Stereo Codec	8051 Uc	PC Parallel Port	Oscillator	Function	UW-FPGA BOARD Pin
1				LSB0			A4					Left LED segment; RAM address line	P35
2				LSB1			A7					Left LED segment; RAM address line	P36
3				LSB2			A5					Left LED segment; RAM address line	P29
4												Uncommitted XS95 I/O pin	
5		DIPSW4						SDOUT	P1.3			DIP switch; codec serial data output; uC I/O	P24
6		DIPSW1				LCEB			P1.0			DIP switch; left RAM chip-enable, uC I/O port	P19
7		DIPSW2				RCEB			P1.1			DIP switch; right RAM chip-enable, uC I/O port	P20
9											CLK	XS Board oscillator	
10			RESETB						XTAL1			Pushbutton; uC clock	P56
11		DIPSW3						MCLK	P1.2			DIP switch; codec master clock; uC I/O port	P23
12												Uncommitted XS95 I/O pin	
13									PSENB			uC program store-enable	
14				S5	RED1							XS Board LED segment; VGA color signal	
15				S6	HSYNCB							XS Board LED segment; VGA horiz. sync.	
17				S3	GREEN1							XS Board LED segment; VGA color signal	
18				S4	RED0							XS Board LED segment; VGA color signal	
19				S2	GREEN0							XS Board LED segment; VGA color signal	
20									ALEB			uC address-latch-enable	
21				S0	BLUE0							XS Board LED segment; VGA color signal	
23				S1	BLUE1							XS Board LED segment; VGA color signal	
25												Uncommitted XS95 I/O pin	
26						KB_CLK			P3.4 (T0)			PS/2 keyboard clock; uC I/O port	
28												JTAG TDI; DIN	
29												JTAG TMS	
30												JTAG TCK; CCLK	
31									P3.0 (RXD)			uC I/O port	
32									P3.7 (RD_)			uC I/O port	
33									P3.5 (T1)			uC I/O port	
34			RDPB						P2.7			Right LED decimal-point; RAM address line; uC I/O port	P41
35				DB8			D7		P0.7			LED; RAM data line; uC muxed address/data line	P61
36				DB7			D6		P0.6			LED; RAM data line; uC muxed address/data line	P62
37				DB6			D5		P0.5			LED; RAM data line; uC muxed address/data line	P65
39				DB5			D4		P0.4			LED; RAM data line; uC muxed address/data line	P66
40				DB4			D3		P0.3			LED; RAM data line; uC muxed address/data line	P57
41				DB3			D2		P0.2			LED; RAM data line; uC muxed address/data line	P58
43				DB2			D1		P0.1			LED; RAM data line; uC muxed address/data line	P59
44				DB1			D0		P0.0			LED; RAM data line; uC muxed address/data line	P60
45									RST			uC reset	
46								CCLK		PC_D0		Codec control line; PC parallel port data output	
47								CDIN		PC_D1		Codec control line; PC parallel port data output	
48								CSB		PC_D2		Codec control line; PC parallel port data output	
49	GND											Power supply ground	
50										PC_D3		PC parallel port data output	
51										PC_D4		PC parallel port data output	
52										PC_D5		PC parallel port data output	
53			RSB4				A12		P2.4			Right LED segment; RAM address line; uC I/O port	P48
54			RSB2				A10		P2.2			Right LED segment; RAM address line; uC I/O port	P45
55			RSB3				A11		P2.3			Right LED segment; RAM address line; uC I/O port	P51
56			RSB1				A9		P2.1			Right LED segment; RAM address line; uC I/O port	P47
57			RSB5				A13		P2.5			Right LED segment; RAM address line; uC I/O port	P50
58			RSB0				A8		P2.0			Right LED segment; RAM address line; uC I/O port	P46
59												JTAG TDO; DOUT	
61			RSB6				A14		P2.6			Right LED segment; RAM address line; uC I/O port	P49
62							OEB					RAM output-enable	
63							WEB		P3.6 (WR_)			RAM write-enable; uC I/O port	
65							CEB					XS Board RAM chip-enable	
66		DIPSW7						LRCK	P1.6	PC_S5		DIP switch; codec left-right channel select; uC I/O port; PC	P27
68									P3.3 (INT1_)			uC I/O port	
69									P3.2 (INT0_)			uC I/O port	
70		DIPSW8				KB_DATA			P3.1 (TX	PC_S6		DIP switch; PS/2 keyboard serial data; uC I/O port; PC par	P28
71		DIPSW6						SDIN	P1.5	PC_S3		DIP switch; codec serial input data; uC I/O port; PC paralle	P26
72		DIPSW5						SCLK	P1.4	PC_S4		DIP switch; codec serial clock; uC I/O port; PC parallel po	P25
74												Uncommitted XS95 I/O pin	
75				LSB3			A0					Left LED segment; RAM address line	P44
76												Uncommitted XS95 I/O pin	
77												Uncommitted XS95 I/O pin	
78	+5V											+5V power source	
79				LSB4			A1					Left LED segment; RAM address line	P38
80										PC_D7		PC parallel port data output	
81										PC_D6		PC parallel port data output	
82				LSB5			A2					Left LED segment; RAM address line	P40
83				LSB6			A6					Left LED segment; RAM address line	P39
84				LDPB			A3					Left LED decimal-point; RAM address line	P37
24,67			SPARE	DP	VSYNCB				P1.7			Pushbutton; XS Board LED decimal-point; VGA horiz. syn	P18

XStend V1.3 - Switches, LEDs, I/O

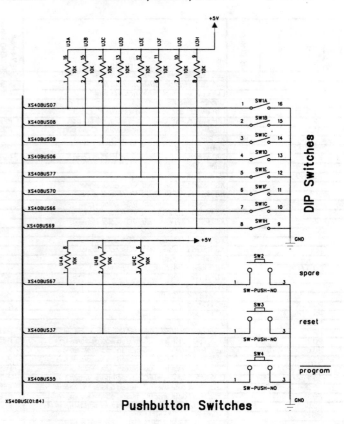

DIP Switches

Pushbutton Switches

XS40BUS[01:84]

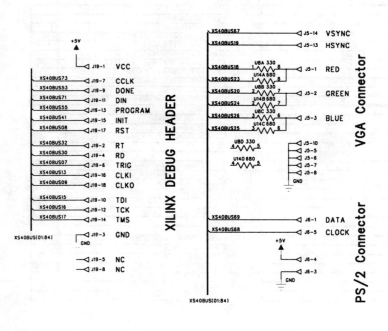

XILINX DEBUG HEADER

VGA Connector

PS/2 Connector

XS40BUS[01:84]

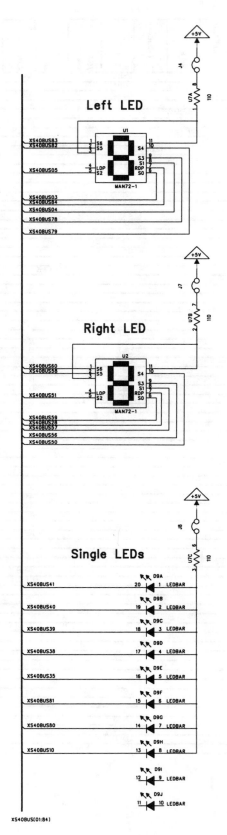

Left LED

Right LED

Single LEDs

XS40BUS[01:84]

1.0 INTRODUCTION

The RSR® Electronics PLDT-1® digital logic trainer board, shown in *Figure 1*, has been designed as a "target board" for students and other users to design, implement, and test digital circuits using a modern programmable device and industry-standard design tools. It is built around the XILINX® corporation XC95108 CPLD device in an 84-pin PLCC package.

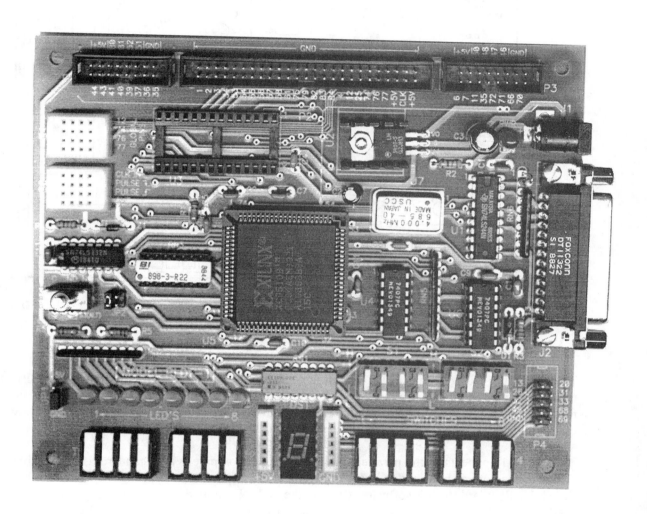

Figure 1

PLDT-1

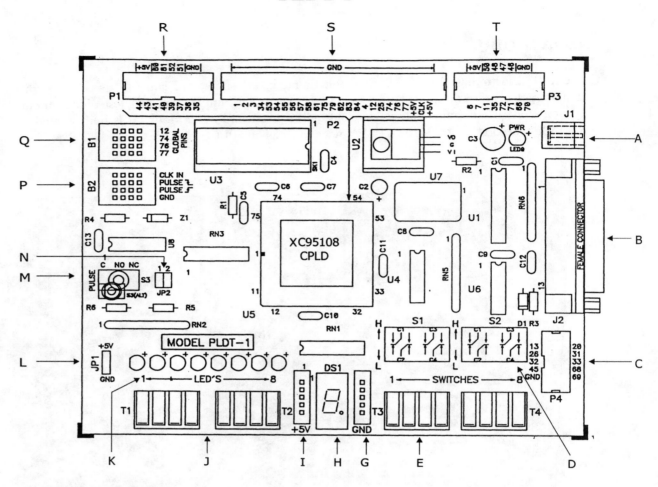

GENERAL DESCRIPTION

The XC95108™ used in the PLDT-1™ digital logic trainer is in-service programmable (ISP). ISP means that the CPLD can be erased and reprogrammed while it is in a circuit, so a separate device-programmer box is not required. The PLDT-1 is powered by a wall-mounted power module which comes with the trainer.

The PLDT-1 connects to the parallel port of a personal computer via a standard 25-wire cable with DB25 connectors at both ends (*Detail B*). The cable is part of the package when purchasing the PLDT-1. Standard design software, such as the XILINX Foundation Series 1.5 Software, can be used with the PLDT-1. Once commands are down-loaded from the PC to the CPLD, the cable may be disconnected; the CPLD "remembers" the design.

Eight toggle switches and eight LEDs can be accessed via screwless terminal blocks (*Details E, J*) to allow connection of external circuits. Also on board the PLDT-1 is a momentary push-button switches, a buffered clock input, a 4 MHz clock oscillator, and a 7-segment display.

NOTES

NOTES

NOTES